ALSO BY ALFRED J. LIMA

❦

The Taunton Heritage River Guide: A Guide to the Appreication of the Taunton River at Fall River (2002)

❦

Preserving Community Character: A Citizen's Guide to Saving Place and Halting Urban Sprawl (2009)

❦

An Oral History of Fall River (tentative title) (2014)

A RIVER AND ITS CITY

The influence of the Quequechan River on the
development of Fall River, Massachusetts

ALFRED J. LIMA

With contributions by Kenneth M. Champlin
and Everett J. Castro

SECOND EDITION

Fall River, MA

PearTree Press
P.O. Box 9585
Fall River, MA 02720
peartree-press.com
riveranditscity.com

All rights reserved. No part of this book may be used or reproduced in any manner whatsoever scanned, or distributed in any printed or electronic form without written permission from the publisher, except in the case of brief quotations embodied in critical articles and reviews.

For information, write us at PearTree Press, P.O. Box 9585, Fall River, MA 02720.

LIBRARY OF CONGRESS CONTROL NUMBER: 2013957003

ISBN-10: 0-9819043-7-8
ISBN-13: 978-0-9819043-7-5

Printed in the United States of America on acid-free paper.

Book design by Stefani Koorey

Copyright © 2007, 2013 by Alfred J. Lima
Chapter 12 copyright © 2007, 2013 by Kenneth M. Champlin
"Quequechan Flora and Fauna: Then and Now" copyright © 2007, 2013 by Everett J. Castro

Excerpts from *Colonial Craftsmen and the Beginnings of American Industry* appear subject to copyright © 1965 by Edwin Tunis. Copyright renewed 1993 by David Hutton. Originally published by World Publishing Company, reprinted 1976 by Thomas Y. Crowell. Reprinted by permission of Curtis Brown, Ltd.
Excerpts from *King Philip's War: The History and Legacy of America's Forgotten Conflict,* copyright © 1999 by Eric B. Schulz and Michael J. Tougias are reprinted with permission of the publisher, The Countryman Press/W.W. Norton & Company, Inc.

Cover images: 1877 map of Fall River, photographed by Michael Brimbau
Lewis Wickes Hine, "Spooler tender—American Linen Co." 1916, Library of Congress

Printed in the United States of America

A RIVER AND ITS CITY

CONTENTS

List of Illustrations	xi
Introduction	xvii
1. The Ecology of the Quequechan River	1
Landscape Characteristics of the Quequechan River Watershed	1
Quequechan Flora and Fauna: Then and Now	5
2. Native Tribes Along the Quequechan River and Surrounding Area	11
Tribal Territories	11
Tribal Use of the Land and the Rivers	15
Tribal Travel Routes	21
3. Early Legends and Explorers of the Fall River Area	23
The Vikings	23
The Legend of the "Skeleton in Armor"	24
The Legends of Dighton Rock	29
Verrazano the Explorer	29
Pirates and Privateers in the Fall River Area	30
The Legend of the "Crone of the Quequechan"	31
4. King Philip's War	39
5. The Colonial Era in Freetown	45
The Town's Early Economy	45
Steep Brook in the Colonial Era	51
Fall River in the Colonial Era	52
6. Colonial Industries on the Quequechan River	55
The Saw Mills	55
The Grist Mills	57
The Fulling Mills	59
The Blacksmith	60

Shipbuilding	62
Whaling in Fall River	63
Coopering	63
Colonial Textile Production	65
Tanneries	66
Shoe Production	68
Salt Production	68
Wood Harvesting	69
7. Early Iron Making on the Quequechan River	73
The Importance of Iron in the Early American Colonies	74
The Elements Required to Manufacture Iron	75
Colonial Iron Manufacturing	78
The Beginnings of Iron Manufacturing on the Quequechan River	82
8. The American Revolutionary Battle on the Quequechan River	85
9. Fall River's Growth Into a Major Textile Manufacturing Center	89
Fall River's Development Within the Context of the Industrial Revolution	89
The Introduction of Textile Manufacturing into America	96
The Rhode Island and Waltham Textile Manufacturing Systems	99
Early Cotton Textile Manufacturing on the Quequechan River	100
Early Textile Printing on the Quequechan River	108
Fall River's Advantage as a Textile Center	110
The Basic Processes of Cotton Textile Manufacturing	113
Power and the Textile Industry in Fall River	122
The Maturing Textile Industry in Fall River	127
The Importance of Textile Machinery	128
Riparian Rights to the Quequechan River	133
10. The Architecture and Evolution of Fall River's Textile Mills	137
The Four Periods of Mill Development in Fall River	137
The Geographic Distribution of Granite and Brick Mills in Fall River	153
How Architecture Reflected the Division of Work in the Mills	154
The Creators of the Mills: Architects, Engineers, Builders, and Their Clients	156
Preserving the Architectural Integrity of the Remaining Mills	160
11. The Fall River Iron Works and its Subsidiary Enterprises	163
The Fall River Iron Works	163
The American Printing Company	170

Shipbuilding Activity of the Iron Works	175
The Iron Works and its Local Steamboat Service	176
The Iron Works and its Railroads	179
The Fall River Line	182
12. The Granite Quarries of Fall River	191
The Italians in Fall River	196
13. The Underground Railroad in Fall River	203
14. The People of the Quequechan River Valley	217
Working Conditions in the Early Mills	217
Housing of Mill Operatives	222
Wages and the Company Store	223
Labor Unrest and Improvement in Working Conditions	225
Environmental Pollution, Working Conditions, and Public Health	229
The Immigrant Culture	230
15. The Environmental Impact of Industrialization on the Quequechan River	235
Historical Sources of the River's Pollution	235
Past Efforts to Remediate Pollution in the River	238
Current Efforts to Clean Up the River	239
16. The Future of the Quequechan River	241
References	249

LIST OF ILLUSTRATIONS

Figure		Page
1.1	Topographic map of the city of Fall River	2
1.2	The three environments of the Quequechan River	3
1.3	Tributary ponds of the Quequechan River	4
1.4	The Quequechan River falls in 1948	6
1.5	The Quequechan River falls today at the Metacomet Mill	7
2.1	Algonquin tribes of Massachusetts	12
2.2	Contact period core areas	14
2.3	Examples of artifacts found at Peace Haven site by Roy Athern	16
2.4	Long houses, typical winter dwellings of the Algonquin	17
2.5	Wampanoag canoe passage	19
2.6	Contact period native trails	20
3.1	Supposed routes of the Viking voyagers	24
3.2	Longfellow's "Skeleton in Armor"	28
4.1	The early battles of King Philip's War and Philip's escape route through Fall River	41
5.1	Typical American schooner in the Colonial Era	47
5.2	Colonial period core areas	49
5.3	Federal period core areas	50
5.4	The West End in 1812	54
6.1	Sawyers sawing a log in a saw pit	56
6.2	Workings of a typical gang saw powered by water on the Quequechan River	57
6.3	The workings of a typical New England grist mill	58
6.4	The mechanism operating thumpers in a fulling mill	59
6.5	Illustration of a fuller at work	60

6.6	The blacksmith shop	61
6.7	A typical New England shipyard	62
6.8	Cooper assembling the staves of a barrel	64
6.9	Home textile production in Colonial America	65
6.10	The tanyard	67
6.11	Construction of a wooden ship	71
7.1	The making of charcoal	77
7.2	An early smelting furnace or bloomery	79
7.3	Operation of a bellows at a blast furnace	79
7.4	A water wheel operating a trip hammer	80
7.5	A water wheel operating bellows at a finery forge	81
8.1	The Lafayette-Durfee house, 94 Cherry Street	87
9.1	The spinning jenny	92
9.2	Horrocks power loom	93
9.3	The early steam engines	95
9.4	Warp and weft	97
9.5	Moses Brown	98
9.6	Samuel Slater	98
9.7	The Colonial Joseph Durfee Mill at Globe Corners	101
9.8	David Anthony	102
9.9	The Arkwright spinning frame	104
9.10	Calico printing, about 1836	106
9.11	Fall River Four Corners, 1843	107
9.12	Commerce on the Taunton and Quequechan Rivers, 1877	109
9.13	The Quequechan River valley, about 1877	111
9.14	Location of textile mills in Fall River, 1912	112
9.15a	How raw cotton is transformed into thread in a textile mill	114
9.15b	How raw cotton is transformed into thread in a textile mill	115
9.15c	How raw cotton is transformed into thread in a textile mill	115
9.15d	How raw cotton is transformed into thread in a textile mill	116
9.15e	How raw cotton is transformed into thread in a textile mill	116
9.15f	How raw cotton is transformed into thread in a textile mill	117
9.15g	How raw cotton is transformed into thread in a textile mill	117
9.15h	How raw cotton is transformed into thread in a textile mill	118
9.16	A breast water wheel	120
9.17	Power transmission in a typical Rhode Island system textile mill	121
9.18	The Watt beam steam engine	123
9.19	The Corliss steam engine	123
9.20	Textile mills located on the Quequechan River, 1915	125

9.21	Mill girl at an early factory loom	126
9.22	Mule spinning	128
9.23	A countershaft with self-aligning bearings	130
9.24	The Fall River Water Works Power Station on North Watuppa Pond	131
10.1	The Metacomet Mill	140
10.2	Oliver Chace Thread Mill	141
10.3	Oliver Chace Thread Mill office building	141
10.4	Union Mill No. 1	143
10.5	The Durfee Mill complex	143
10.6	Tecumseh Mill No. 1	144
10.7	The Crescent Mill	144
10.8	The Chace Mill bell tower	145
10.9	The Chace Mill	145
10.10	Italian Romanesque campaniles	146
10.11	Mill towers in the third phase of mill development	146
10.12	The Arkwright Mill and Davis Mill No. 2	148
10.13	Davol Mill No. 2: A classic example of mills in the fourth phase	148
10.14	The Durfee Mills weaving shed	150
10.15	The Seaconnet Mills	150
10.16	Characteristic mill chimney	151
10.17	Mill workmanship	151
10.18	Mechanics Mill	152
10.19	Davol Mills	155
10.20	The Merchants Mill	157
10.21	The Border City Mills	158
10.22	Elements integral to the preservation of the architectural integrity of Fall River's mill buildings	161
11.1	American Print Works and the Fall River Iron Works, 1877	165
11.2	Colonel Richard Borden	166
11.3	Jefferson Borden	167
11.4	A sample of print patterns from the American Print Works	169
11.5	American Printing Company	171
11.6	M.C.D. Borden starting the Corliss steam engine in Mill No. 4, American Printing Company, 1895	173
11.7	Matthew Chaloner Durfee Borden	175
11.8	Fall River's waterfront, about 1900	177
11.9	The Fall River Line steamer *Bristol*	181
11.10	The New Line boat, Fall River	182
11.11	Rail lines serving the Fall River Line steamers	183
11.12	The Fall River Line steamer *Commonwealth*	185

11.13	The Fall River Line steamer *Providence*	187
11.14	The freight steamer *City of Fall River*	188
11.15	The steamboat *Puritan*	189
12.1	Fall River city hall engraving	193
12.2	Commercial granite quarry sites in the Quequechan River valley	194
12.3	Early photographs of quarrying operations at the Beattie quarry	197
12.4	The Rolling Rock on the edge of the quarry in its original setting	201
12.5	Early engraving of the Rolling Rock	201
13.1	Arnold Buffum, anti-slavery leader, in 1826	205
13.2	Officers of the New England Anti-Slavery Society	206
13.3	An incident described in the *Anti-Slavery Record*	208
13.4	Samuel G. Chace and Elizabeth Buffum Chace	209
13.5	Nathaniel Briggs Borden	210
13.6	Known Underground Railroad way stations in Fall River	212
13.7	A flag made by Arnold Buffum Chace	213
13.8	Routes of the Underground Railroad in Eastern Massachusetts and Rhode Island	215
14.1	Bell schedule for a New England textile mill, 1853	221
14.2	Typical layout of a Rhode Island system mill village	222
14.3	Petition of Fall River mill workers for a 10-hour work day	224
14.4	The "bread strike" of 1875	225
14.5	Young operatives in the Cornell Mill, Fall River, 1912	227
14.6	The Fall River dinner pail	231
14.7	The progression of urban growth in Fall River	232
14.8	Downtown Fall River, about 1890	233
15.1	Privies overhanging river above Watuppa Dam, October 2, 1914	236
15.2	Choate Street sewer; crude sewage discharged into river, October 2, 1914	236
15.3	Privies, south end of Blossom's Avenue, October 2, 1914.	236
15.4	3:15 pm, September 22, 1909, from 100 feet below Fall River Laundry, looking down stream	237
15.5	Privy at discharge from Hargraves Mills Nos. 2 and 3, October 2, 1914	237
15.6	City dump between Lawrence and Salem Streets, October 13, 1914	237
16.1	Recommendations of the Fall River Urban River Visions Initiative	242

A RIVER AND ITS CITY

INTRODUCTION

For such a relatively small stream, the history and impact of the Quequechan River on the Southeast region of Massachusetts, the United States, and the world is rather remarkable.

No other river can boast of all of the following: it has unique geologic characteristics; native tribes have used it for thousands of years; it has been the location of the full gamut of Colonial industries; it was the site of one of the significant battles of King Philip's War; it is the backdrop of a poem by Henry Wadsworth Longfellow; it has its own witch legend; and it was the site of a battle with the British during the Revolutionary War. In addition, it provided the power that initiated the textile industry in Fall River, and it later provided the process water for cooling the steam engines for textile mills, making the city the largest cotton textile manufacturing center in the country at the time and the largest in the world after Manchester, England.

In the Quequechan's small valley there existed a thriving granite quarrying industry and along the river the conductors of the Underground Railroad ferried freed slaves to safety. In addition, the river's lower falls powered both a major iron manufacturing center and a shipbuilding industry. At the river's mouth, whaling ships moored and whale oil processed. The home pier of the great Fall River Line steamers was located where the river enters Mount Hope Bay. It is no exaggeration to say that the whole history of the northeastern United States, from earliest native tribal communities to the end of the industrial era, can be told through this river.

The Quequechan, however, is also an example of the environmental toll imposed on rivers during the Industrial Revolution and of the effort to undo that damage and make the river a focus of urban revitalization. Undoubtedly, without the Quequechan River, there would be no Fall River as we know it today. Even the name of the city is taken from the river's falls.

This book's purpose is to relate how the Quequechan River influenced the development of Fall River, from its beginnings as a Colonial village, through its emergence as a great cotton textile manufacturing center.

Alfred J. Lima

Please note: *A River and Its City* is too comprehensive to provide a detailed analysis of any one topic. Throughout the text, the reader is directed to the many sources used. Footnotes have two numbers, for example (45-34). The first number designates the source in the References section, and the second number is the page on which you can find that information.

Please visit us at **RiverAndItsCity.com** and share your memories, family stories, and thoughts on this great river.

1 The Ecology of the Quequechan River

Landscape Characteristics of the Quequechan River Watershed

Bedrock: In Fall River, the landscape of the Quequechan River is determined by underlying bedrock, which is principally granite. Granite was formed billions of years ago by the cooling of molten lava that exuded from the earth's core, then uplifted over millions of years. The high elevations of these bedrock formations provide Fall River's dramatic topography and the views of the Taunton River and Mount Hope Bay.

Unlike the flat surrounding coastal landscape, the uplifted bedrock provides Fall River with commanding views to the west. The elevation of the city ranges from sea level at the Taunton River on the west side of the city to 354 feet above sea level in the eastern part of the city at Copicut Hill.

Unlike most communities in the United States, bedrock became an important consideration in Fall River's economy. Granite quarries were common in the city and local granite was used in constructing all of the city's attractive granite textile mills. Fall River granite was also used in constructing and dressing public and private buildings in this city and in other communities.

fig. 1.1 **Topographic map of the city of Fall River**
Contour intervals are 20 feet.
Source: *Fall River 2000*.

This steep granite uplift has created three different environments of the Quequechan River that have been critical to the growth of the city.

Above the hill is the fresh water environment that includes the upland, wetlands, ponds, and streams that all contribute to the flow of the Quequechan River. This fresh water environment or "watershed" is an expansive area of 34 square miles that extends into East Fall River and beyond the city's boundaries into Westport and Tiverton, Rhode Island. The damming of the Quequechan River above the falls increased the storage capacity of the watershed and created an extensive area of flat standing water. The visual effect of the river here is one of several placid ponds.

The hillside is the second environment of the Quequechan, where fresh water meets steep topography. The placid water above the hillside now becomes a raging torrent, rushing 130 feet down the steep gradient. The falls gave the river its name, which the native tribes called "Falling Water." The combination of the gradient change and water volume results in the creation of latent power.

The third environment of the Quequechan River is where it meets navigable salt water below the hill at the point where the Taunton River enters Mount Hope Bay. Unlike the area above the hill, the landscape at the mouth of the river is relatively small in area. Typically,

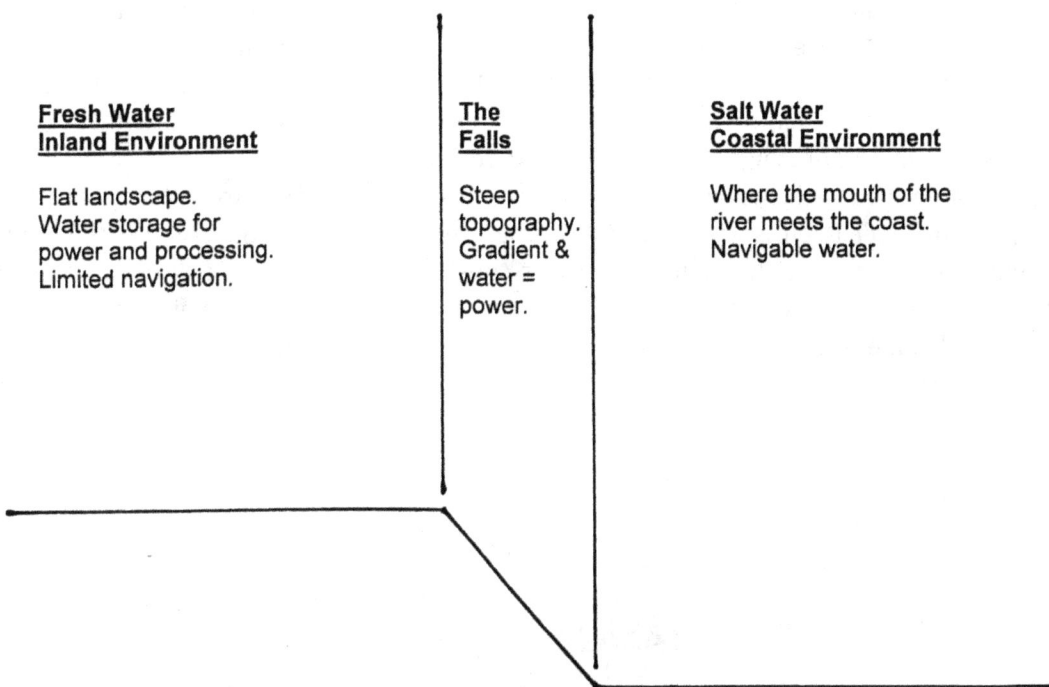

fig. 1.2 **The three environments of The Quequechan River**

the falls of a fresh water river would be considerably inland from a coastal area and not navigable to the coast. However, in Fall River, the Quequechan falls are right at the coast, a factor that would prove influential in the development of the city.

All three of these landscapes have been essential to the development of the city of Fall River. The environment above the hill has created the volume of fresh water required for textile processing. The hillside environment provided the power for the early growth of the textile industry. The lower salt-water environment provided immediate water access to the outside world, allowing the efficient delivery of bulk raw goods to the city and finished goods to be shipped away. The landscape of Fall River has shaped its history in a manner that is unique among communities.

The Quequechan River: Quequechan is a Algonkian word meaning "Falling Water." The main characteristic of the Quequechan River is its dramatic drop within a relatively short distance. The river falls approximately 130 feet within a distance of 2,300 feet. Its watershed includes the Watuppa Pond, Stafford Pond, Sawdy Pond, and Devol Pond.

In common with most naturally-occurring ponds in the northeast, these five ponds originated as "kettle holes." Kettle holes were formed during the waning of the last Ice Age when large ice blocks broke away from the receding glaciers. Over many years, the surrounding glacial lake deposited sand and gravel around the ice blocks. When the glacial lake drained and the ice blocks melted, they left water-filled indentations, or ponds. A well-known kettle hole is Walden Pond in Concord, Massachusetts. The map titled "Tributary Ponds of the Quequechan River" (fig. 1.3) shows the five tributary ponds with a strong north/south axis, indicating the direction of the receding glaciers.

The watershed of the Quequechan River is approximately 27 square miles in area. The Quequechan River itself is only two miles long, from the head of the river at the South Watuppa Pond outlet to the dam. A dam was constructed at the top of the falls at Third Street that raised the level of the feeder ponds by five feet. The river has eight falls during the last half-mile of its course as it descends over its granite bed to tidewater at Mount Hope Bay. (30-39)*

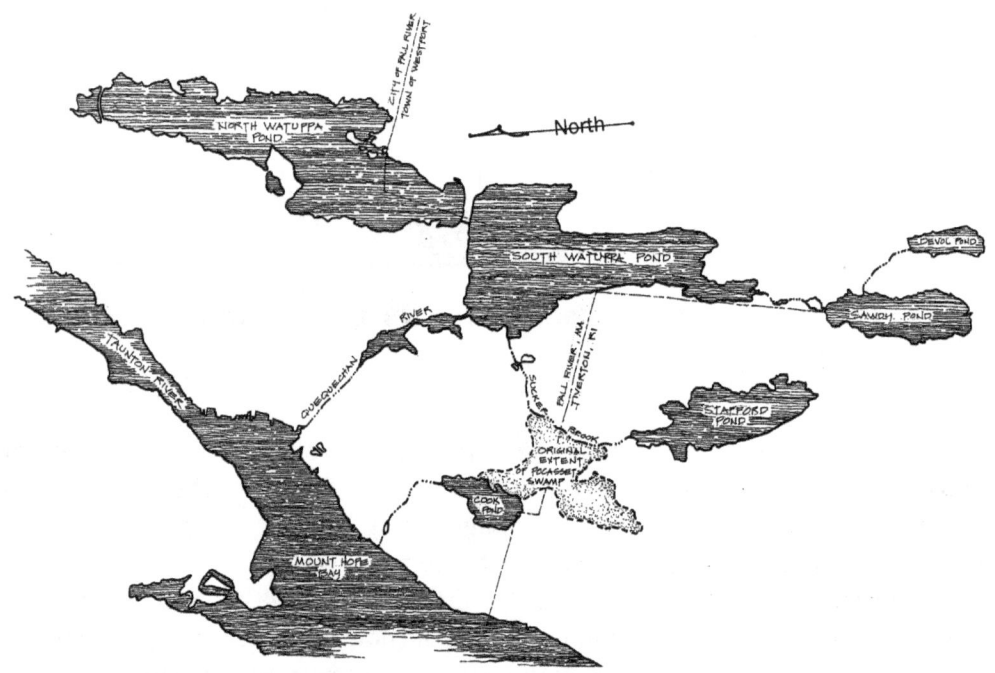

fig. 1.3 **Tributary ponds of the Quequechan River**

---

* (30-39): This citation signifies that the source of the information is number 30 in the Reference section. The number 39 identifies the page from which the reference is taken. Therefore, the information cited was taken from page 39 of Fay, Spofford, and Thorndike, *Report of the Watuppa Ponds and Quequechan River Commission to the City Council, City of Fall River*.

Quequechan Flora and Fauna: Then and Now
by Everett J. Castro

Today, it is difficult to imagine the beauty and bounty of the Quequechan River valley prior to European settlement. What did the Quequechan River look like? What species of plants and animals frequented its shores and lived in its waters?

The length of the river was, of course, the same as we see it today, but its width, both above and below the falls, changed markedly with the seasons. The Quequechan's flood plain extended from Ruggles Park in the north to beyond Rodman Street to the south and the river's width varied seasonally depending on the amount of water flowing from its source ponds. Once dammed, a number of granite ridges just south and west of where the Route 24, Brayton Avenue, and the Route I-195 highway interchanges are today became islands. These islands slowly disappeared, year by year, as the coves and floodplain of the Quequechan were filled for commercial and industrial development or used as a handy repository for the growing city's domestic and industrial waste.

Unlike the relatively static surface area of the Quequechan we see today, the pristine Quequechan's water area would have been much more dynamic and water flow down the Quequechan would have been much greater. When Watuppa Pond, the main source of the Quequechan and the second largest natural water body in Massachusetts (Assawompsett Pond, a few miles to the northeast in Lakeville, being the largest), was separated at the Narrows into two distinct bodies of water, all of the water lying to the north, now North Watuppa Pond, became the municipal water supply. North Watuppa water was no longer available to supply the Quequechan. The Quequechan's flow was now solely dependent on the watershed of the now South Watuppa and the adjoining ponds, Stafford, Sawdy, and Devol.

History tells us that there were many Indian settlements along the upper Quequechan and it is easy to see why. For semi-nomadic hunter-gatherers, the short two mile length of the river not only meant that all areas of the river's ecosystem were easily accessible by foot, but that those two miles also provided an amazing variety of wildlife habitats in a small space. From the freshwater Watuppa source to saltwater marshes and tidal estuary at its mouth, the Quequechan served as a "supermarket" to the indigenous Wampanoag.

The Quequechan River and its feeder ponds provided a source of fresh water to the natives as well as pickerel, yellow perch, pumpkinseed sunfish, white suckers, and brown bullhead catfish. Tributary streams would have contained colorful brook trout all year round and, in the spring, white suckers would make their spawning runs to spawn over upstream gravel beds.

Interestingly, fish found in the Quequechan and feeder ponds today are much more varied than when the Wampanoag fished these waters. It wasn't long after European settlement that westward-moving pioneers discovered new fish species and began to transport fish eggs and fry hither and yon seeking to augment or "improve" on

fig. 1.4 **The Quequechan River falls in 1948**

The cascading waters of the Queqeuchan River at the site of the old Pocasset Mill, opposite the Fall River Herald News building off of Pocasset Street. This photo was taken in 1948, before the river was placed in a conduit in the 1960's as part of the construction of Route 1-195.

nature. Fish were one of the first alien wildlife species to arrive in this area via the hand of man.

Largemouth bass, the major freshwater sport fish found in the Quequechan and Watuppa today, were brought here from the Midwest. The feisty smallmouth bass came from New York State. Other now abundant alien species found in the Watuppa and Quequechan include the bluegill sunfish, black crappie, and various pike/muskellunge crosses.

If you see someone fishing in the Quequechan today, chances are they are hoping to catch a largemouth bass.

Brook trout, intolerant of degraded environments and eutrophic conditions, are no longer present. Brook trout require cold, well-oxygenated water to thrive and reproduce.

Although the Quequechan's 130' falls would have prevented salt-water anadromous fish species (fish that move into fresh water to spawn) from ascending the river, the area below the falls would have provided spawning sites for saltwater stream spawning species such as blueback herring and rainbow smelt. Despite the degraded environment in that area, a remnant population of river herring still spawn every spring in the Quequechan

fig. 1.5 **The Quequechan River falls today at the Metacomet Mill**
During a winter low-flow period. Photo by Al Lima.

between the Central Street Bridge and where the river emerges at the railroad tracks under the Braga Bridge.

Atlantic salmon, Atlantic sturgeon, striped bass, and other saltwater species abundant in the pre-European Taunton River could not ascend the steep falls and so were not found in the Quequechan.

At the Quequechan's mouth and along the adjacent tidal shoreline and salt marshes of the Taunton River would have been vast beds of oysters, soft shell clams, blue and ribbed mussels, blue crabs, and northern quahogs.

As for waterfowl, the upper ponds and Quequechan shallows would have provided huge beds of wild rice and wapatoo, or "duck potato." These plants would not only have been extremely attractive to migrating waterfowl, but also would also have been eagerly gathered and eaten by the local human population. Canada geese, black ducks, green and blue-wing teal, redheads, and canvasbacks would have been trapped and netted by folks living along the river during the spring and fall.

When the fresh waters of the Watuppa and Quequechan froze, the turbulence below the falls and the freeze retardant salt water at the mouth of the Quequechan would have attracted hardy waterfowl, such as Atlantic brant, during the cold winter months. The Quequechan was a four-season resource pantry, providing fish and fowl somewhere along its short two-mile length throughout the year.

Even today, the Quequechan holds an amazing variety of waterfowl during migration periods. Canada geese are still extremely abundant. Green Wing Teal appear in the fall. Ruddy ducks and hooded mergansers are seen along the Quequechan in late fall and early spring. Unfortunately, some species have not done as well. The shy black duck has yielded to its relative, the more human-tolerant, cosmopolitan urban mallard. Other native waterfowl have been displaced by the large alien mute swan.

Although beautiful to view as they glide about the Quequechan in the heart of the city, mute swans are extremely territorial and ferocious in driving lesser waterfowl off what the swans appear to consider their territory. This swan dominance limits native waterfowl species diversity on the Quequechan and other area waters.

Fur was another Quequechan resource. Furs were needed to survive the harsh New England winter. Raccoon, mink, muskrat, otter, and beaver supplied fur clothing and bedding. With the exception of the beaver, all can still be found living along the Quequechan today.

Even now, there are still older residents of the Flint section of Fall River, and individuals from other Quequechan neighborhoods, who recall making enough money to get by on by trapping muskrats and other furbearers in the cattail marshes of the Quequechan during the Great Depression of the 1930s.

One individual recounted how he used to go with his grandfather, every winter, to the Borden Mill, adjacent to the Quequechan, to ask permission of Spencer Borden to trap muskrats along the river on mill property.

Red fox, skunks, weasels, white tail deer, and black bears were abundant in the uplands adjacent to the Quequechan. Although not as plentiful today, and with the exception of the bear, they can all still be found almost within the center of the city.

The vegetation along the river is vastly different from what it was originally. Huge beds of cattails and wild rice have given way to impenetrable masses of common Phragmites. Phragmites tolerate contaminated soil and thrive in polluted water. Unfortunately, unlike the native cattail, they have little value as a food resource for wildlife.

The beautiful alien purple loosestrife, that can be seen blooming every summer along the river as one travels along Route I-195, is another pernicious invasive driving out more delicate native species. Norway maple is an alien similarly invading riverside areas.

Aquatic alien plants such as milfoil, fanwort, and South American waterweed are flourishing in the river. These invasive aquatic weeds thrive in enriched polluted environments. Native species require cleaner water and have been out-competed by these hardier introductions that have probably arrived via discarded home aquarium plants.

Although vastly changed from its pristine past to its urban present, the Quequechan River today still contains an amazing array of native and non-native species of flora and fauna.

A RIVER AND ITS CITY

For a river running through the heart of a large Massachusetts' city, and for a river that has been so exploited and abused by humans, it is amazing that one can still see such a diversity of life in its limited green areas and waters. Every summer, ospreys still circle the upper river on the lookout for a fish dinner, and peregrine falcons, as they have for thousands of years, patrol the airspace over the mouth of the river where it flows into the Taunton.

2 Native Tribes Along the Quequechan River and Surrounding Area

The cultural history of the Quequechan River is closely tied to its geologic history. Approximately 12,000 years ago, following the re-vegetation of the area once the glaciers receded, nomadic tribes of hunter-gatherers migrated into New England. In northern New England, these tribes continued to be hunter-gatherers; however, in southern New England, the hunter-gatherer tribes also began—with the introduction into the northeast of maize (corn), squash, and beans about 1,000 AD—to cultivate fields to supplement their food supply.

Tribal Territories

These early hunter-gatherers eventually evolved into a linguistic family of tribal nations that ranged over most of the northeastern United States, known as the Algonquin. In New England, there were seven principal confederated Algonquin nations, including the Abenaki (Maine), the Penacook (New Hampshire), Pequot (Connecticut), the Narragansett (Rhode Island), the Pocumtuck, the Nipmuck, the Massachusetts, and the Wampanoag. The map on

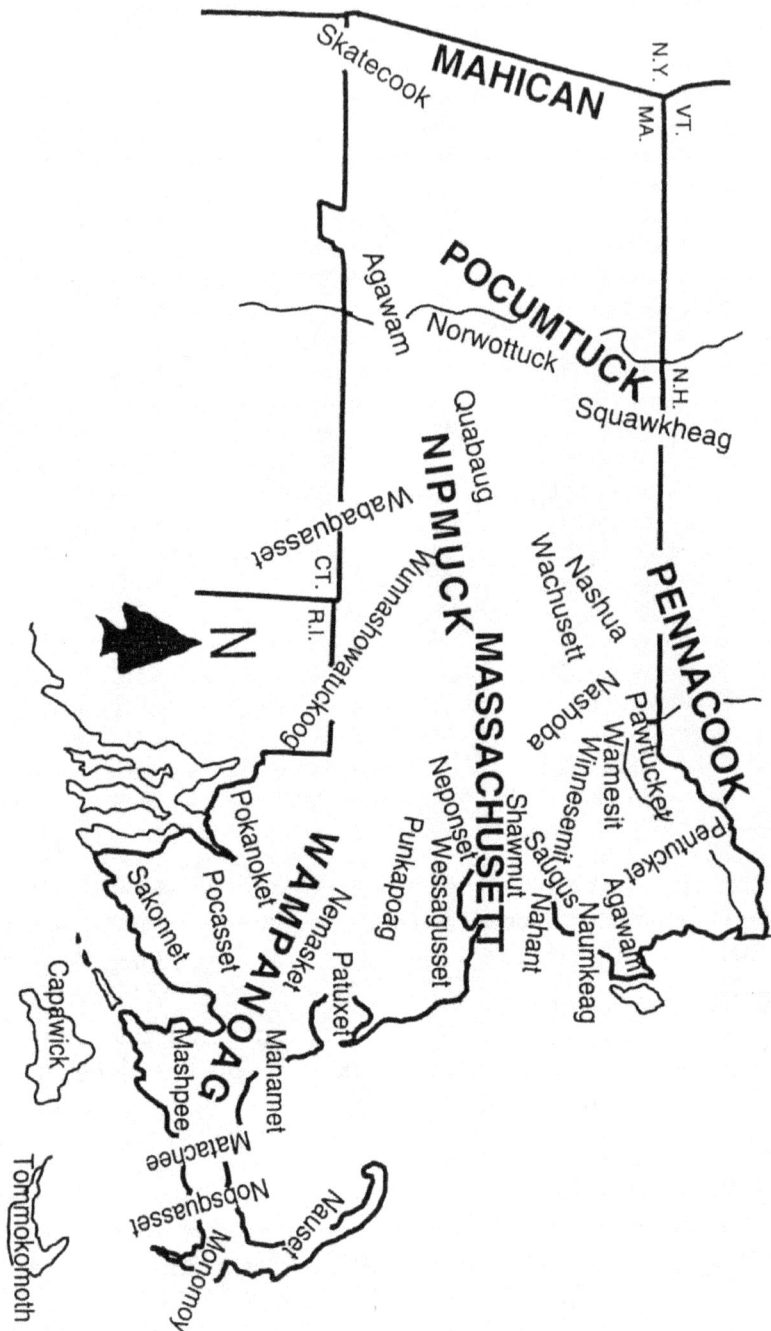

fig. 2.1 **Algonquin tribes of Massachusetts**
Source: Johnson: Ninnouck: The Algonkian People of New England.

the previous page shows the former territories of these nations and their tribes.

The tribal confederacy of the Wampanoag wielded power from the eastern shore of Narragansett Bay to Cape Cod and the islands. The tribes of the Wampanoag nation included the Pokanoket, the lead tribe of the Wampanoag. Their territory included what is now Bristol, Warren, Barrington, Rehoboth, Seekonk, and Swansea, except for Gardner's Neck. The Pocasset were the next most powerful of the Wampanoag. The Sakonnet were based in Little Compton. The Nemasket territory included the upper reaches of the Taunton River. Along the lower reaches of the Taunton River, the dominant native tribe was the Pocasset, whose territory straddled the Taunton River and extended from Gardner's Neck on the west to the Westport town line on the east, from the southerly boundary of Tiverton on the south to all of Freetown on the north. It included the present-day towns of Fall River, Somerset, the Gardner's Neck section of Swansea, Tiverton, and Freetown. (63-14) The principal village of the Pocasset was Mattapoisett, or Shawomet in today's Somerset. The map on the following page shows the concentration of native tribal settlements along the Taunton River corridor.

In his *History of Fall River*, Arthur Phillips says that, in the 1670s, the main camp of Weetamoe, the squaw sachem of the Pocasset, was located on the banks of the Quequechan River, probably near Hartwell Street, where the skeleton in armor was found, or at the top of the falls where a spring existed. Several other camp locations on the Taunton River are definitively known. These include locations near the Brightman Street Bridge, at the base of the Quequechan River (now occupied by the Tillotson Company), along the shore of South Watuppa Pond near its outlet into the Quequechan River, in the vicinity of Ruggles Park (then on the shores of the Quequechan River) where, at one time, there was a sizable spring in the hillside (in his *History of the Town of Somerset*, William A. Hart mentions that the Indians did not drink from surface streams but from springs), and at the Wigwam lot next to the Valentine house on North Main Street. (65-9)

During a visit to Mount Hope Bay in 1524 for King Francis I of France, the Florentine explorer Giovanni de Verrazano described the natives who approached his ship:

> They were dressed in deer-skins, wrought with branches like damaske, their hair was tied up behind with divers knots. This is the goodliest people, and of the fairest conditions that we have found in this our voyage; they exceed us in bigness, they are of the color of brass, some of them incline more to whiteness, other are of yellow color, of comely visage, with long and black hair, which they are very careful to trim and deck up, they are of sweet and pleasant countenance. The women are very handsome and well formed, of pleasant countenance, and comely to behold; they are as well-mannered as any women, they wear deer skins branched embroidered as the men use, there are also some of them which wear on their arms very rich skins of Luzernes [weasels], they wear divers ornaments according to the usage of the people of the East. (61-25)

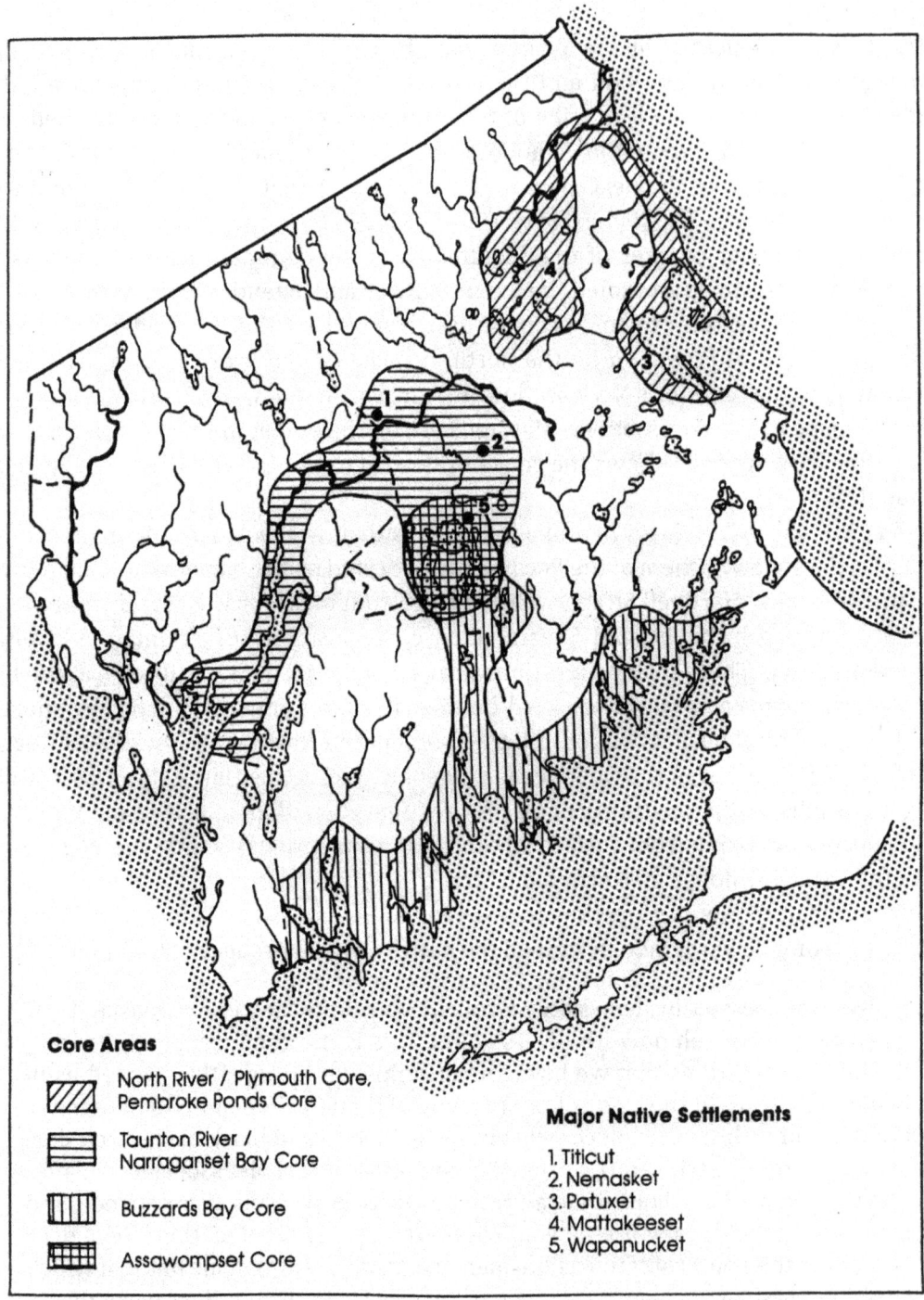

fig. 2.2 **Contact period core areas**

Source: Massachusetts Historical Commission: *Historic and Archaeological Resources of Southeast Massachusetts.*

Tribal Use of the Land and the Rivers

The Algonquin were most populous in southern New England. Here, the native tribes did not passively live off of the land but actively managed it to maximize its productivity. These tribes practiced both an extensive and an intensive form of husbandry. The extensive form of land management involved the yearly burning of woodland understory growth, which provided easier travel, more efficient hunting, allowed lush grass growth that supported larger wildlife populations, and reduced diseases and pests. The saltwater marshes were also burned annually to maximize meadow grass growth and therefore encourage wildfowl productivity.

According to one early historical account:

> From early times the Indians has been accustomed to burn over the whole country annually in November, after the leaves had fallen and the grass had become dry, which kept the meadows [marshes] clean, and prevented any growth of underbrush on the uplands. One by one the older trees would give way, and thus many cleared fields, or tracts with only here and there a tree, would abound, where the sod would be friable, ready for the plow; or already be well covered with grass, ready for pasturage. The meadow lands thus burnt over, threw out an early and rich growth of nutritious grasses, which if let alone, grew 'up to a man's face.' (9-8)

The Pocasset avoided the dense, silty clay soils of glacial lake bottoms because they were more difficult to work and because they tended to remain wet longer into the spring season. They also avoided the stony till soils of the upland areas (prevalent in most of Fall River), because of their difficulty to cultivate, and the upper kame terrace soils (in Assonet) because of their dryness.

The soils most prized by the Pocasset were those low-stage kame terrace and kame delta deposits prevalent along much of the Taunton River in Swansea, Somerset, and Freetown. These soils were neither too wet nor too dry, easily workable with simple tools and with adequate topsoil. The most valuable and commonly used agricultural fields were those near tributary streams where fish weirs could be constructed to catch anadromous fish that swam upstream to spawn in the spring. One of the most used of these areas was at Peace Haven at the mouth of Mother's Brook.

Salt marshes are productive ecosystems and native tribes often located near these environments. The Taunton River and its tributary streams provided a wide range of fresh and saltwater fish, shellfish (soft shelled clams, quahogs, oysters, periwinkles), and crustaceans (crabs and lobsters). All of these were brewed into a chowder called nasaump. (44-21) The woods provided abundant game in season, and the woods and marshes along the river were habitats for a many species of wildfowl. The burning of the woodlands and grasslands resulted in abundant berry production, including blueberries, strawberries,

Hole stones or perforated weights found at the Peace Haven site. The function of these stones has not been definitively determined, but they are believed to have been used to weigh down nets to catch alewives during Spring spawning season on Mother's Brook. A total of 265 of these stones were found at the Peace Haven site. These stones have been found in only a few sites and in only Southern New England; however, only Peace Haven has yielded so many of these stones. The upper middle stone is 5 inches wide.

A spear head and arrowheads found at the Peace Haven site. The spear head is almost 8 inches long.

fig. 2.3 Examples of artifacts found at Peace Haven site by Roy Athern

Source: Athearn: *Bulletin of the Massachusetts Archaeological Society*

raspberries, grapes, blackberries, elderberries, and huckleberries. Wild cranberries abounded in the local swamps. In addition, trees provided a variety of nuts, cherries, plums, and other fruit. In the spring, maples provided the sap that was boiled down to make maple syrup. Birches provided a thinner but still sweet syrup.

Over their open fires, the squaws broiled roe, boiled succotash, baked cornbread, and refined the sugar of the maples. Winter was a season of semi-starvation, but the fish and meat that was dried in the sun saw the tribe through to the next growing season. Even through the Indians were surrounded by seawater, they never learned the use of salt to preserve meat and fish. (44-21)

Medicinal herbs were readily available to treat a variety of ailments, and there was considerable trade in these commodities.

In addition to fish and shellfish, the Taunton River provided many other necessities of life. One of the most important of these necessities, salt to flavor their food, was derived from river salt water that was placed in stone hollows and evaporated. The thick, hard quahog shells provided blades for hoes; other shells were fashioned into fishhooks and other implements, including small tweezers that young braves used to pull out the hair on their sparse beards. The Taunton River marshes yielded clay for making pottery and provided rushes for basket making and for the mats that the Pocasset wove for summerhouse walls and beds.

The Taunton River even provided currency for the native tribes. Whelk shell and the

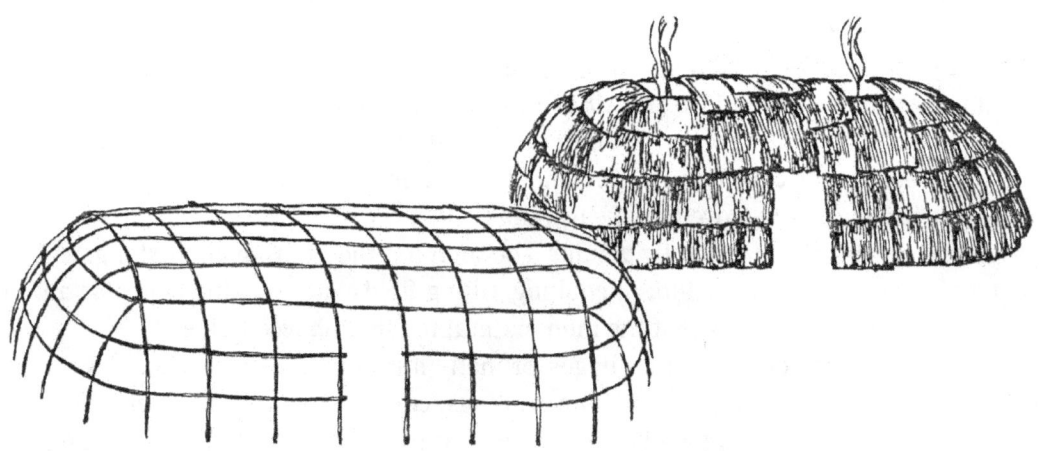

fig. 2.4 **Long houses, typical winter dwellings of the Algonquin**

The framework for the long houses consisted of two parallel rows of saplings set in the ground. They were then bent over and lashed to each other. The ends were set in a semi-circle to join the main framework. The coverings consisted of deerskins or bark, with smoke holes on top. The skins or bark were held down and made weather-proof with a series of horizontal saplings that were secured to the main frame. The entrances had covers made of skins.

Source: Wilbur: *The New England Indians*

white of the quahog shell furnished the material for the valuable white beads of wampum and the purple spot on quahog shells provided the more valuable purple bead. Wampum was used for personal ornamentation, but its high regard among the natives was noticed by Dutch traders, and wampum later became a medium of exchange and acted as currency. When English currency later became scarce, the European settlers used wampum amongst themselves as an equivalent to money. The Narragansett and, later, the Wampanoag learned to make wampum, which originally came from tribes on Long Island. Governor Bradford of the Plymouth Colony noted in his journal that wampum "makes ye Indians in these parts rich & power full and also prowd therby."

The agricultural practices of the Pocasset involved the planting of corn on mounds in the spring, with beans planted at the base to allow the vines to use the corn stalks as a pole. Squash, pumpkins, and a root vegetable, which we know today as Jerusalem artichoke, were planted in between. Alewives were dug into the base of the corn mounds to provide fertilizer. One source mentions that crop rotation was practiced, with one third of a field cultivated, leaving the other two-thirds to recuperate for two years. (29a-11) However, more frequently, the use of fish for fertilizer allowed fields to be used year after year. Large fields were cultivated on a cooperative basis, as the concept of personal land ownership was unknown to native tribes.

John Winthrop noted that the time of planting by the Indians was regulated by natural seasonal events: "Some of the Indians take the time of the coming up of a Fish, called Aloofes [Alewife], into the Rivers. Others of the budding of some Trees." Later commentators said that the proper planting time for corn was "when the leaves of the White Oak are as big as the ear of a mouse." (44-20)

Once the fall harvest was over and festivals concluded, the surplus foodstuffs were placed in storage areas dug into the sides of dry sandy hillsides. The tribe would then divide into small bands and spread out into the woodland for the fall hunting season. Once the hunting season was concluded in December, the band reassembled inland in protected areas where firewood was more readily available.

Division of work was divided by gender. The men did the hunting and fishing, and the women did everything else, including cooking, tilling fields, and making clothing, among other tasks. The exception was that the men maintained the tobacco fields.

Native settlements occurred in villages of many structures. Mild weather dwellings, called wetus, were round and built of thin saplings, covered by reed mats, and held one or two families. In the winter, long houses were built inland and held up to eight families. These long houses were also built of a frame of saplings lashed to one another and covered with deerskins or bark, with smoke holes in the top. The early Europeans found these houses as warm and snug as their frame dwellings. A typical long house was 50 feet in length. Ceremonial long houses could extend up to 100 or 200 feet in length and 30 feet in width. (86-44)

Native tribes were prodigious users of firewood, since they kept home fires going constantly. As a result, the landscape inland along the shoreline of the Taunton River and

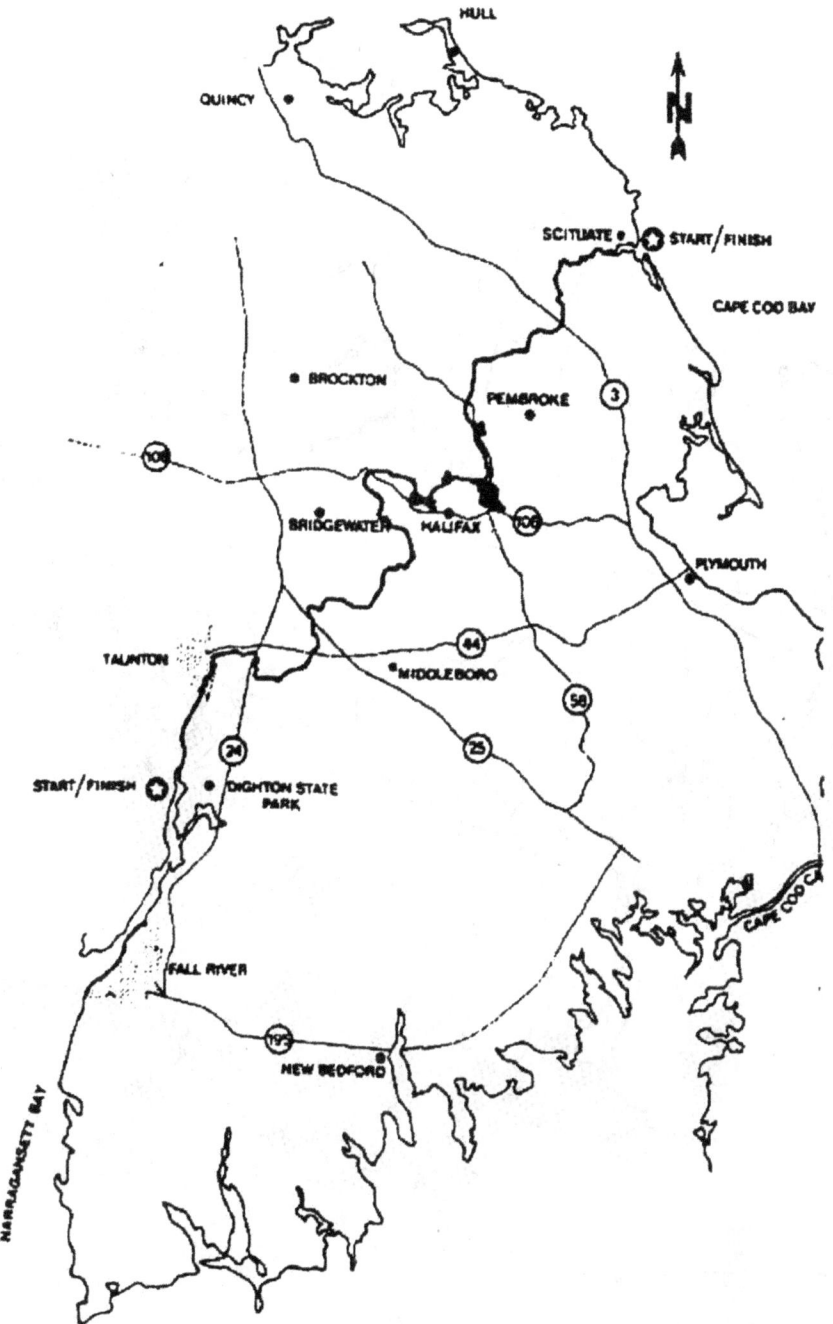

fig. 2.5 **Wampanoag canoe passage**

In traveling from Mount Hope Bay to Massachusetts Bay, the Wampanoag traveled up the Taunton River in dugout logs to the high point and watershed divide at Little Sandy Pond in what is now Pembroke. They then proceeded down the Nemasket River to Massachusetts Bay.

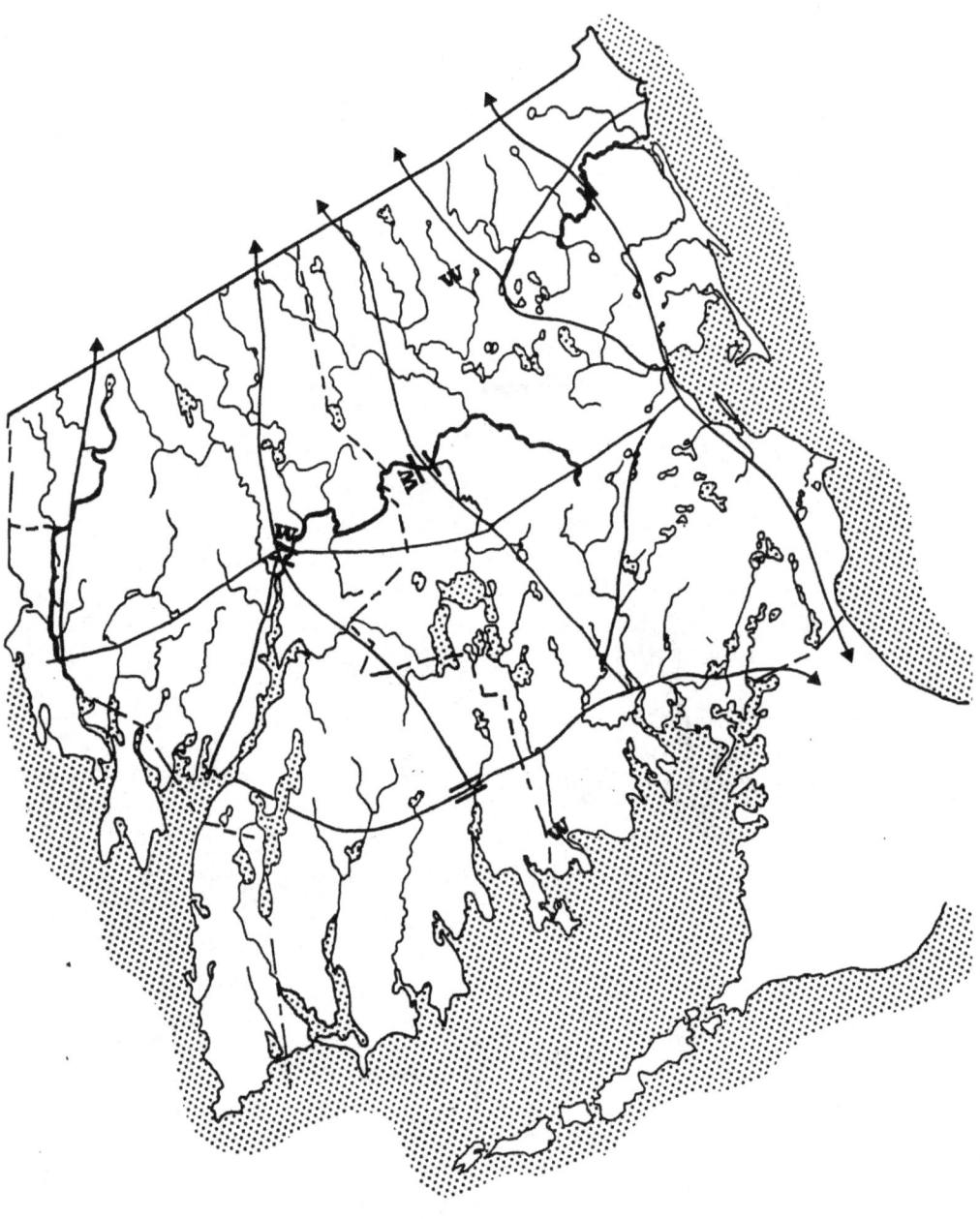

fig. 2.6 **Contact period native trails**

Source: Massachusetts Historical Commission:
Historic and Archaeological Resources of Southeast Massachusetts.

Mount Hope Bay was virtually denuded of forests. Early European explorers of Mount Hope Bay and the Taunton River remarked that the landscape had been cleared for cultivation and pasture for a considerable distance back from the shoreline.

Tribal Travel Routes

The Taunton River was a major route of travel and communication for native tribes. Travel on the river was by dugout logs (made by burning out the center of the log). These log boats could seat as many as 12 persons. The Wampanoag used an overland connection by canoe from Mount Hope Bay to Massachusetts Bay via the Taunton and North Rivers. The high point and watershed divide was at Little Sandy Pond in Pembroke. To the north, a connection was made with the North River and to Massachusetts Bay. This would have been a much safer, quicker and practical route to these water bodies than traveling along the coast. See the map on page 19 for the route of this Wampanoag canoe passage.

A major path led from Taunton south along the western side of the Taunton River to Somerset, paralleling what later became Route 138, and to the tip of Gardner's Neck. At the southern tip of Gardner's Neck, a ferry was operated by Corbitant, chief of the Pocasset when the Pilgrims first arrived, to Mount Hope, seat of the Wampanoag. Another major path used by the Pocasset was the Mowry Path, which proceeded along the eastern side of "Watuppa Lake" (the ponds were once one water body, and a bridge crossing at the Narrows did not occur until 1828). The Mowry Path connected Buzzards Bay with the Taunton River at Peace Haven. There was a tribal ferry at nearby Winslow's Point, where the river narrows. Wamsutta, the son of Massasoit and husband of Weetamoe, reserved an area of land at that location out of the Freetown Purchase in 1657 "to be used by the Indians who kept the ferry." The map titled "Contact Period Native Trails" on the previous page shows native trails that later became Colonial ways.

In 1617, a plague struck the local tribes in the area, probably brought by European traders. The Wampanoag were especially affected and lost at least two-thirds of their population of 3,000. When the Europeans arrived, they found a land that was not virgin, but widowed.

3 Early Legends and Explorers in the Fall River Area

The Vikings

Early European exploration of Mount Hope Bay and its tributary rivers such as the Quequechan is shrouded in romantic legend, beginning with the Vikings. In his *History of the Town of Somerset*, William Hart reported that Norwegian scholars had long believed that Leif Ericson built his "leifbooths" at Mount Hope in the year 1000 and wintered there. Further, that Thorwald, a brother of Leif, was to have spent the years 1002 and 1003 along the coast of Southeastern Massachusetts. Thorfall, another brother, was thought to have sailed in the area in 1007 and named Cape Cod "Wonder Strand," Buzzards Bay "Stream Furth," Martha's Vineyard "Stream Isle," and Mt. Hope "Hop." The map on the following page shows alleged Viking routes to the new world that was included in Phillips' *History of Fall River*.

Historians now believe that Vineland was located in Newfoundland and that there is no specific evidence that Lief Ericson, Thorwald, or Thorfall visited Mount Hope (Hop) at all. Before 1960, it was believed that it was on Mt. Hope that Thorfall's wife bore him a son, Snorre, who would have been the first white child born in the New World. The legend continued that native tribes attacked the Vikings during the next fall and the party removed to Buzzards Bay until Snorre was three. (61-21) After the Vikings left, it would be 500 years before Europeans would again "discover" the New World.

For a while, it was thought that the Vikings inscribed the Dighton Rock with their markings, but that legend has been discredited. However, if the Vikings had made a settlement on Mount Hope Bay, it is very likely that they would have explored up the Taunton River.

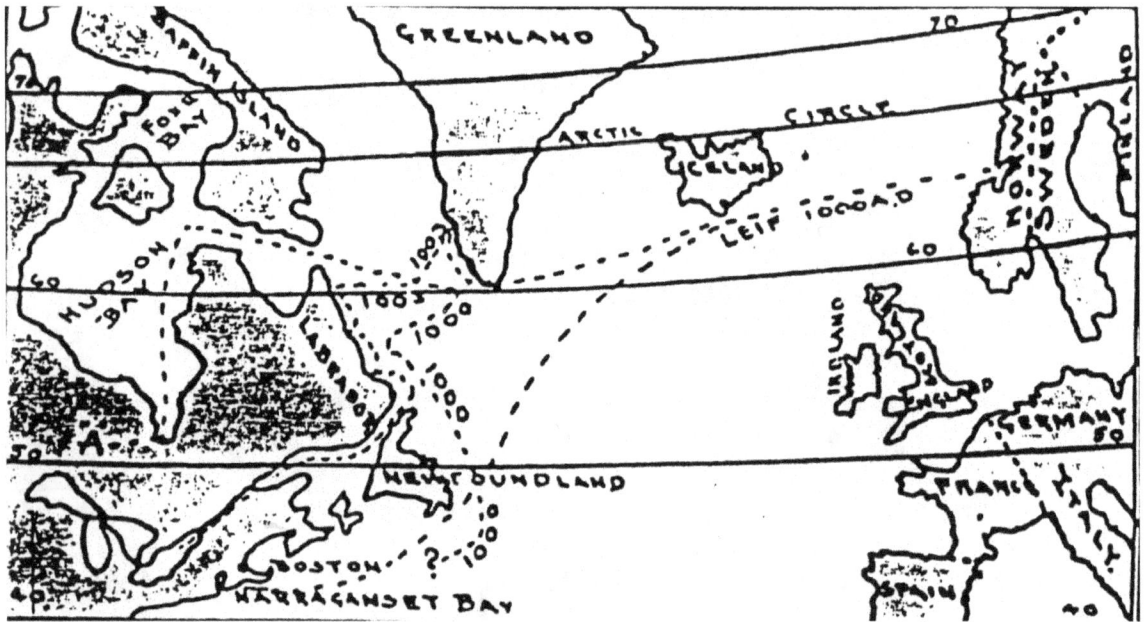

fig. 3.1 **Supposed routes of the Viking voyagers**

Source: Phillips: *History of Fall River*

The Legend of the "Skeleton in Armor"

The Victorians romanticized the Vikings in their art and literature. Romantic poet Henry Wadsworth Longfellow, while visiting his brother Samuel (then minister of the Unitarian Society in Fall River) became intrigued with a skeleton that was found in 1832 in perfect condition in an embankment on the shores of the Quequechan River at Hartwell Street and Fifth Streets and displayed at the Fall River Athenaeum. The skeleton had a triangular brass breastplate, a belt of brass tubes about four or five inches long, and was found near the surface in a sitting position. Arrowheads and parts of other skeletons were found nearby. The sitting skeleton was believed to be that of a chief.

This gave Longfellow the idea to write "The Skeleton in Armor," a poem that linked the skeleton in Fall River and the stone tower at Newport into a story of a rebellious Viking and his maiden who narrowly escape by sea the wrath of her father and eventually establish a new home on the banks of the Quequechan River. Longfellow spent many summers in Newport and became intrigued with the stone tower, which, at that time, was believed built by the Vikings.

Later, less romantic interpreters would point out that the local Pocasset buried their dead in similar armor. The 26-foot-high tower in Newport was later established to have been built as the base of a windmill and is first mentioned in written records in 1677 in the will of Benedict Arnold (grandfather of the infamous traitor). Arnold was President and later Governor of the Rhode Island Colony between 1662 and his death in 1678. The tower is one of the oldest surviving man-made structures in the United States.

But then, no one knows with certainty....

"The Skeleton in the Armor" by Henry Wadsworth Longfellow

"Speak! speak! thou fearful guest!
Who, with thy hollow breast
Still in rude armor drest,
    Comest to daunt me!
Wrapt not in Eastern balms,
But with thy fleshless palms
Stretched, as if asking alms,
    Why dost thou haunt me?"

Then, from those cavernous eyes
Pale flashes seemed to rise,
As when the Northern skies
    Gleam in December;
And, like the water's flow
Under December's snow,
Came a dull voice of woe
    From the heart's chamber.

"I was Viking old!
My deeds, though manifold,
No Skald in song has told,
    No Saga taught thee!
Take heed, that in thy verse
Thou dost the tale rehearse,
Else dread a dead man's curse;
    For this I sought thee.

"Far in the Northern Land,
    By the wild Baltic's strand,
    I, with my childish hand,
        Tamed the gerfalcon;
And with my skates fast-bound,
Skimmed the half-frozen Sound,
That the poor whimpering hound
        Trembled to walk on.

"Oft to his frozen lair
    Tracked I the grisly bear,
    While from my path the hare
        Fled like a shadow;
Oft through the forest dark
Followed the were-wolf's bark,
Until the soaring lark
        Sang from the meadow.

"But when I older grew,
    Joining a corsair's crew,
    O'er the dark sea I flew
        With the marauders.
Wild was the life we led;
Many the souls that sped,
Many the hearts that bled,
        By our stern orders.

Alfred J. Lima

"Many a wassail-bout
 Wore the long Winter out;
 Often our midnight shout
  Set the cocks crowing,
 As we the Berserk's tale
 Measured in cups of ale,
 Draining the oaken pail,
  Filled to o'erflowing.

"Once as I told in glee
 Tales of the stormy sea,
 Soft eyes did gaze on me,
  Burning yet tender;
 And as the white stars shine
 On the dark Norway pine,
 On that dark heart of mine
  Fell their soft splendor.

"I wooed the blue-eyed maid,
 Yielding, yet half afraid,
 And in the forest's shade
  Our vows were plighted.
 Under its loosened vest
 Fluttered her little breast,
 Like birds within their nest
  By the hawk frighted.

"Bright in her father's hall
 Shields gleamed upon the wall,
 Loud sang the minstrels all,
  Chanting his glory;
 When of old Hildebrand
 I asked his daughter's hand,
 Mute did the minstrels stand
  To hear my story.

"While the brown ale he quaffed,
 Loud then the champion laughed,
 And as the wind-gusts waft
  The sea-foam brightly,
 So the loud laugh of scorn,
 Out of those lips unshorn,
 From the deep drinking-horn
  Blew the foam lightly.

"She was a Prince's child,
 I but a Viking wild,
 And though she blushed and smiled,
  I was discarded!
 Should not the dove so white
 Follow the sea-mew's flight,
 Why did they leave that night
  Her nest unguarded?

"Scarce did I put to sea,
 Bearing the maid with me,
 Fairest of all was she
  Among the Norsemen!
 When on the night sea-strand,
 Waving his armed hand,
 Saw we old Hildebrand,
  With twenty horsemen.

"Then launched they to the blast,
 Bent like a reed each mast,
 Yet we were gaining fast,
  When the wind failed us;
 And with a sudden flaw
 Came round the gusty Skaw,
 So that our foe we saw
  Laugh as he hailed us.

A RIVER AND ITS CITY

"And as to catch the gale
  Round veered the flapping sail,
  'Death!' Was the helmsman's hail,
     'Death without quarter!'
Mid-ships with iron keel
Struck we her ribs of steel;
Sown her black hulk did reel
     Through the black water!

"As with his wings aslant,
  Sails the fierce cormorant,
  Seeking some rocky haunt,
     With his prey laden,
So toward the open main,
Beating to sea again,
Through the wild hurricane,
     Bore I the maiden.

"Three weeks we westward bore,
And when the storm was o'er,
Cloud-like we saw the shore
     Stretching to leeward;
There for my lady's bower
Built I the lofty tower,
Which to this very hour,
     Stands looking seaward.

"There lived we many years;
Time dried the maiden's tears;
She had forgot her fears,
     She was a mother;
Death closed her mild blue eyes,
Under that tower she lies;
Never shall the sun arise
     On such another!

"Still grew my bosom then,
  Still as a stagnant fen!
  Hateful to me were men,
     The sunlight hateful!
In the vast forest here,
Clad in my warlike gear,
Fell I upon my spear,
     O, death was grateful!

"Thus, seamed with many scars,
  Bursting these prison bars,
  Up to its native stars
     My soul ascended!
There from the flowing bowl
Deep drinks the warrior's soul,
Skoal! to the Northland! skoal!"
     Thus the tale ended.

SKELETON IN ARMOR

In his poem, "The Skeleton in the Armor," Longfellow connected the tales of Viking visits to Mount Hope Bay with the armored skeleton found in Fall River and the stone tower in Newport, once thought to have been built by the Vikings.

fig. 3.2 **Longfellow's Skeleton in Armor**

The Legends of Dighton Rock

The first recorded European explorers of Narragansett Bay were John and Sebastian Cabot in 1497, although it is uncertain whether they entered Mount Hope Bay or the Taunton River.

In 1502, a Portuguese fishing captain and explorer named Miguel de Corte-Real set sail from Europe to look for his brother Gaspar, who had not returned from a trip to the New World. However, Miguel, too, was never heard from again. Legend has it that Miguel de Corte-Real at one time entered the Taunton River at Fall River and lived among the Pocasset. The only clue that we have of his life here is an cryptic inscription on a rock in Berkley (Dighton Rock) that some have interpreted as saying in Latin "M. Cortereal 1511 V. Dei Dux Ind." Loosely translated, this abbreviation says "by the grace of God, Leader of the Indians." This implies that Corte-Real lived among the Indians more than a century before the Pilgrims. Was he, or did he fancy himself to be, a leader of the natives? We shall never know.

Other inscriptions on Dighton Rock have led scholars in recent times to theorize that Phoenicians, Vikings, and even pirates like Captain Kidd and Blackbeard might have carved inscriptions into the rock, providing directions to buried treasure. No less than 34 theories have been put forward on the origins of the inscriptions on the rock. Why this rock in particular should have been a well used "guest book" of sorts for historical personalities is itself a mystery.

Verrazano the Explorer

In 1524, Giovanni di Verrazano of Florence, who was exploring the New World for King Francis I of France, entered Narragansett Bay and Mount Hope Bay. He continued up the Taunton River past Fall River, probably to Dighton, where he noted in his report to the king that a ship could sail up the river's course and that thereafter it might be navigated by "small shallops."

It was Verrazano who remarked in his journal that the landscape along Mt. Hope Bay and the Taunton River was not all wooded, as we would expect, but "park-like" in appearance, with lush grassland and occasional large trees rising a considerable distance from the shoreline. This was a result of the extensive use of firewood for fuel by native tribes for thousands of years, which kept their home fires burning continuously year-round. It was also the function of the annual burning of woodlands to encourage the growth of grassland for food for wild animals that they hunted and for ease of travel through the woodlands. Since the local tribes lived near to the shoreline, the area near to the shore would, over time, become denuded of trees and indeed be park-like in appearance.

By the time that Bartholomew Gosnold arrived in the area in 1602, he found that Dutch, English, and French trading vessels had become frequent visitors to Mount Hope, the seat of the Wampanoag and a favorite trading point for them.

Alfred J. Lima

Pirates and Privateers in the Fall River Area

During the Colonial period, New England privateers would capture and plunder ships of nations that were at war with England, including at various times ships of French, Dutch, and Spanish origin. This eventually led to outright piracy in many instances, with Newport known to be a base of privateers and a few pirates. In the Quequechan River legend, the "Crone of the Quequechan," a mysterious witch-like person was feared by early settlers because of her suspected powers and her past association as mistress to pirate Captain Kidd, although Black Sam Bellamy, a Newport-based pirate, might have been the real nefarious seaman.

Governments actually initiated the practice of organized piracy in the Caribbean. During the mid 1600s, the English home government—and virtually every Colonial governor—commissioned local sea captains to become privateers to prey on Spanish shipping in the West Indies. Any country that happened to be at war with England at the time—Spain, France, the Netherlands, and Portugal—was fair game. Sometimes, the delay in communication in those days would result in unfortunate consequences. After plundering Portuguese commerce in 1704, Captain John Quelch returned to Marblehead to find that England had signed a peace treaty with Portugal. The unfortunate privateer Quelch was promptly hanged in Boston as a pirate. (22-220)

The pirate's plunder could be substantial. Pirate Captain Black Sam Bellamy filled the galley of the ship *Whydah* with plunder from months of piracy. The astonishing fortune included 30,000 pounds of silver, 10,000 pounds of gold, 20 tons of ivory, and "enough precious jewels to ransom a princess," according to a crew member. (50) While carrying this booty, the *Whydah* was wrecked and sank during a severe storm off of Cape Cod near Wellfleet in the spring of 1717. A total of 142 crewmembers were drowned. The six pirates who made it to shore were tried in Boston and hanged on a scaffold erected between "the ebb and flow of the tide." (22-221)

The practice of executing pirates between low and high tide seems to have evolved through time-honored custom and by an Act of Parliament, which decreed that the gallows should be erected "in such place on the sea, or within the ebbing and flowing thereof, as the President of the Court ... shall appoint." These spectacles were usually attended by hundreds of citizens, who came either by land or observed from their boats. (22-223)

The treaty of peace between England and Spain, signed in 1668, caused a great many privateers, now unemployed, to turn to pirating in the West Indies and along the New England coast. To the destitute seaman, there was little difference between the lawfulness of the privateer and the unlawfulness of the pirate.

Newport was a hub for pirates and profited well from them. One Peterson arrived in a ship of ten guns and 70 men to be refitted at Newport. A grand jury was convened, but no bill could be made against him because many members of the grand jury were neighbors or friends of the crew. A Robert Munday was also tried for piracy in Newport in 1703. (22-146) Privateers probably also originated from ports along the Taunton River, particularly from

Assonet and Somerset, but no history book records their exploits.

The gains from privateering and piracy were very large, with profits of 2,900 percent not unusual. Pirates usually did not threaten the source of their income, and took their cargo and left with the crew unharmed. Often the cargo found its way back to the port of origin, and instances of Boston merchants in connivance with pirates were not uncommon. (22-148)

Before the English navy reduced piracy in the Caribbean around 1715, there were a recorded 1,500 pirates preying the seas around the West Indies.

It was commonly believed that the story of the *Whydah* was a legend but, in 1985, after several years of effort, Cape Cod underwater explorer Barry Clifford found the remains of the ship off of Wellfleet. His team recovered millions of dollars of treasure from the remains of the *Whydah*, including artifacts, coins, and gold dust. Some of the ship's artifacts are on display in a small museum of pirate history in Provincetown. (56)

The Legend of "The Crone of the Quequechan"

The tale of the Crone of the Quequechan first appeared in print in the April 3, 1845 issue of The Fall River *Weekly News*. The origins of the legend are unknown:

"Has the flesh and blood a charm against heated iron and boiling oil?" Scott asks.

On the southern bank of the Quequechan, the Indian name for the little stream which passes through Fall River, there stood near the middle of the 18th century, a miserable looking log hut, perched upon a rocky eminence overhanging the great falls, not far from where now stands the Massasoit mill. It was constructed of unhewn logs, unskillfully locked together at their angular joinings, and occupied a spot close upon the brink of the chasm through which the current below dashed and foamed in its passage to the bay. The building has a one-sided antique appearance. Its roof, in places, had partially fallen in and one of its corners, that nearest to the margin of the stream, had settled away, so as to incline the structure over the yawning chasm, into which it seemed ready to tumble at any moment without aid of any extraordinary force.

By whom it was created, and why in that uninviting, if not dangerous spot, nobody could tell. It had long been deserted except by bats and reptiles, and was fast going to decay under the alternate action of the sun and rain.

Site Elevated

The ground or rock on which the building stood was the most elevated spot on the southern bank of the stream, and overlooked the whole surrounding country, including the opposite shore of Mount Hope Bay, except the southeastern portion which was covered by higher land in that direction.

Like most deserted houses it had a bad name. Lights, it was said, had been seen on dark and stormy nights, through the crevices in the walls, sometimes emitting red or blue flames;

and strange noises had more than once been heard by passers-by, issuing from this dilapidated structure.

Many, indeed, were the stories of what had been seen and heard about the premises, and few among the inhabitants were hardy enough to trouble its vicinity much if it could be avoided after nightfall. Such was the general dread in which the building was held, that the inhabitants had more than once threatened to burn it down, to shove it over the brink into the foaming flood below, but the threat, like many others, had never been put in execution, and so the hut stood, year after year, the theme of many a wild and thrilling tale.

Light Seen in Hut

On a cold bleak night in the month of December, when the good quiet people in the neighborhood of the Quequechan had sought their smugly closed rooms, well heated with great, burning logs, around which young and old were gathered in the full consciousness of comfort and security, a brilliant light was seen pouring through the holes and crevices of the deserted hut, as though its interior were lighted up by innumerable torches; and what had never been before seen, a fiery column of sparks was observed to escape from the top of the little chimney, indicating that a blaze was raging on its ancient hearth.

By not a few, light had been seen there before, and frightful stories told of the sudden appearance of an old woman, strangely habited, about the hut on such occasions, known among he inhabitants generally as the "Crone of the Quequechan." But here was a brilliant illumination long before the usual hour of retiring, conspicuous to the eyes of the whole neighborhood.

The fact was undeniable: its intensity attracted the observation of the whole population, and the first movement was a gathering of all ages and sexes at the grocery store of Peter Leonard, which stood on the east side of what is now North Main Street, near the spot occupied by the Mount Hope House.

Excitement Heightens

Great was the excitement among the neighbors when assembled. Women and children were crying; strong men looked grave and doubtingly at one another; and all, except Stephen Davis, exhibited unequivocal signs of fear and anxiety.

Stephen was a tall, lank, bony fellow, some two or three and twenty years of age—the noisiest braggart of the settlement, and at all times more ready to pick a quarrel than willing to stand his ground when met by a fair opponent. In the present instance, he was unusually loquacious—called on any three to join him and assist at once to demolish the illuminated hut—said he did not care the value of a cast iron jack knife if Old Scratch and all his family were at the bottom of the matter; all he wanted was enough associated with him, to bear witness of the manner in which he should stir up the spirits. No one regarded the ranting of poor Stephen, for his blustering on all occasions was well known to all the inhabitants. At length, Peter Leonard suggested that something should be done besides talking.

What to Do?

"I propose," he said, "that we take the Bible and move in a body upon the spot."

"I propose," said Lot Lee, "that we first send for the minister at Assonet."

"I propose," said Welcome Brownell, "that we take guns and other weapons of offense and at once attack the hut."

"I propose," said Joel Wilson, "that we stay where we are."

"And I propose, " said Mark Woodard, "that a committee be appointed to visit the premises."

"I second the motion," said Stephen Davis.

"Send four or six men to the spot, and if anything less than a legion of infernal spirits are there, we'll plunge them into the Quequechan."

Crone Appears

This observation was scarcely uttered before the door of the shop flew open, and exhibited standing within the entrance a hideous looking old crone, bent half double with age, her face disfigured with irregular streaks of smut and her blood-shot eyes flashing fire as she gazed on the motley group within.

Her head was thrust forward, exhibiting a nose of uncommon magnitude, covered with warts and carbuncles, beneath which a mouth, half open, extending almost from ear to ear, showed here and there a few long dark tusks projecting out like half burnt stumps in newly cleared field. Her chin was covered with several huge tufts of long gray hair, portions of which curled up and lodged within the corner of her capacious mouth. Her eyebrows were black and shaggy and her ears projected out at right angles with the head, exhibiting the appearance of large curled and crisped pieces of dingy sole-leather.

Her large bony hands, foul with sore and accumulated filth, were thrust forward, and her long hooked fingers, incessantly in motion, seemed eager to seize whoever or whatever might come in her way.

Strange Dress

On her head she wore a scarlet colored handkerchief, the ends of which were brought round and tied in a bow directly in the center of her forehead. Her dress was a scarlet robe, of ample dimensions, which fell down nearly to her ankles, and was tied about her middle with a broad sash of the same color. Her shoes were of cowhide, ornamented with strange figures in red paint, and fastened over the instep with large black cord set off with small red tassels.

As the door flew open the crone placed one foot on the threshold, and leaning forward scanned, for an instant, the several persons within. Then stepping boldly into the room, and contracting her features to a hideous scowl, she exclaimed in tones harsh and grating:

Crone Speaks

"Who talks of throwing me or mine into the Quequechan? Who talks of priests and Bibles? Who of guns and fire?"

Then bursting into a low, hoarse laugh, she fastened her blood-shot eyes on Stephen Davis and, shaking her long bony finger as she spoke, she continued in tones of biting scorn:

"Thou braggart! Thou, Stephen Davis, talk of stirring up spirits? Thou, who dare not enter a cellar in the daytime, talk of firing my hut? I would like to see thee alone, within a rod of its entrance when I am there. Out upon thee, thou scorn of thy race!"

David shrank away under this bold invective, pale as a ghost, and sought protection behind the counter, where stood Lot Lee and Peter Leopard. Not a word was uttered in reply, but here stood the crone, motionless as a statue, still keeping her fiery eyes fixed on Davis as though she would annihilate him by a look.

She Taunts Group

"All dumb!" at length said the crone, looking round the room and scowling horribly, as her eyes fell on each individual of the group in turn. "All dumb, are ye, my masters? O ye are a valiant band, truly! Burn my hut, will ye? No, no, the man is not here that dare do that!"

And she strode up and down the center of the open space, with a firm, heavy trend, looking alternately to the right and left, as she passed between knots of individuals on each side of the room.

She made four or five turns, each time approaching near and nearer the place occupied by Welcome Brownell in front of the great fireplace. At length she came closer to his side and cast at him a quick, fiery look of defiance and indignation. With the quickness of thought Brownell drew back, raised his arm and planted a deadly blow, with his heavy iron fist, directly between the eyes on the lower edge of her forehead.

Crone is Felled

The old crone fell like a log at full length upon the floor, trembling and quivering for a moment in every fiber and muscle of her prostate body. A few seconds elapsed and all was still as death. Her features assumed a livid hue, her breathing became imperceptible, her muscles relaxed, her arms fell heavily down by her sides and she appeared to have paid the last debt of nature without a struggle.

There was the stillness of the tomb in that assembly for several minutes while gazing on the relaxed and motionless body of the old crone. Fear and consternation were depicted on the countenances of the whole group. Not a breathing was heard nor a word uttered. At length Brownell stepping forward lifted one of her arms from the floor; then letting go of it, it fell heavily at her side.

"Merciful God," he exclaimed, stepping back with a look of inexpressible horror, "I have killed the woman."

She Revives

These words were scarcely uttered when the muscles of the crone's face were observed slightly to move; her limbs contracted; her lips parted, her eyes flew open, and in an instant after, she uttered a long, loud howl and sprang to her feet with the easy bound of a tiger!

Terrible as was the presence of the hag, her sudden appearance in life was a joyous relief to the assembled neighbors. Murder was not at all in their hearts, and perhaps least of all in that of poor Brownell, who a moment before, would have given empires could he have withdrawn the blow. It was all right and over now and he laughed outright with excess of joy.

"Laugh, fiend or hell," shouted the crone, casting on Brownell a look of terrible scorn.

Crone Seizes Child

"Laugh imp of the fiery world! But know ye that my revenge is certain and speedy" and away she bounded towards the door, seizing on her passage a child, little more than a year old, sitting in the lap of Mrs. Brownell. One moment more and she was dashing through the narrow street toward the illuminated hut with the speed of a race horse. The mother uttered a piercing shriek and bolted through the doorway, followed by the whole assembly.

Joel Wilson led off, but it was seen at a glance, fleet of foot that he was, that the old crone was more than a match for the young settler, even encumbered as she was with the weight of the child. She dashed over the narrow plank bridge which crossed the stream, near where now stands the Quequechan factory, and mounted the opposite bank with the fleetness of a young deer. Another bound and she stood upon the edge of the rock overhanging the cataract, at the corner of the still illuminated hut. Those in pursuit, one after another, in quick succession, reached the elevated ground a few rods higher up the stream, all conceiving, at one and the same instant, as if by divine communication, the terrible purpose of the old crone.

Threatens Child' Life

"Not another foot nearer!" shouted the infuriated hag in loud and grating tones, holding the shrieking child at arms length over the dashing current below. "Not another foot nearer, ye devils incarnate!—one step and the heir of Brownell never breathes again!" and she made a motion indicating her purpose should her command be disregarded. Awe-struck and palsied stood the whole assembly. No one moved; no one uttered a word. Desperation marked the expression of the terrible being before them.

The mother sank upon the ground, senseless as the clod on which she lay, while the father stood gazing on his child in an attitude of speechless despair.

"Ho! Ho!" at length shouted the crone, breaking out into her accustomed low, hoarse, guttural laugh, and casting on the silent and motionless group a wild and fiendish look of triumph. "Ho, ho, ye dastards; fire my hut, will ye! Ay, fire it! But know, ye knaves while ye destroy with fire the crone of the Quequechan destroys with water" and she bent forward as if in the act of dashing from her the still shrieking child.

While the awe-struck spectators each expected every moment to see put in execution her terrible threat, the athletic form of Lot Lee was seen cautiously but rapidly gliding across the open plot of ground in the rear of the seething hut, keeping that object between him and the fiery hag. As her back was turned in that direction she did not observe his silent approach, and he passed unobserved to the northern angle of the building.

Child is Rescued

From this aspect, a moment after, Lee made a single bound forward and, seizing the crone with one hand, and the child with the other, hurled her with the strength of a giant several feet from the fearful edge of the chasm. As she fell, he leaped towards the spot and placing one foot upon the breast of the prostrate hag, held up the exhausted child freed from the iron grasp of the raging crone!

A long loud shout from the assembled crowd and a sudden rush to the spot followed this unexpected feat of relief. A circle of all ages and sexes was instantly formed around the struggling crone, all eager, as if by instinct, to do the bidding of her athletic conqueror.

"Give me a strong cord," said Lot Lee, still keeping one foot on the hag's chest. "She has no more liberty on these grounds."

A rope was soon procured with which her hands and feet were securely tied, when the pressure on her chest was removed and she was left untouched upon the ground.

"To the hut now," shouted Lee, making towards the dilapidated structure, followed by the whole neighborhood: "Let us see the nest of the she-devil before covereing it into ashes and smoke." And he seized the crone by the shoulders and dragged her towards the entrance of the building.

On trying the door it was found securely fastened, apparently on the inside, and by means not readily overcome. An attempt was made to force it open by a violent push, but a second trial showed that it could not, in this manner, be effected. A low, chuckling laugh was at his moment heard issuing from the throat of the old crone as she lay upon the ground at the foot of Lot Lee.

"Ye cannot do it, ye cowards!" she exclaimed, looking her conqueror full in the face, "mortal strength alone will never open that door, so batter it down if enter ye will."

Hut Door Forced Open

"A stone there!" shouted Lot, speaking to the crowd in the rear. "A stone Stephen; now is your time to stir up the spirits."

A heavy rock was instantly applied to the woody obstruction, which after two or three smart blows, began to yield and soon tumbled down upon the earthy floor within. Lee dragged the crone over the threshold into the apartment, followed by the whole assembly. What a void was there!

The whole building was lighted up by a great fire, made of old stumps and large pine knots, which burned brightly and strongly on its ancient hearth. A solitary stool stood in the center of the floorless apartment, rickety with age, and mouldy from lack of use. Not another article of furniture was anywhere to be seen; not a table or chair or other object except in one corner was a heap of straw on which the hag must long have lain for it was much broken up and matted together as through years had elapsed since first placed there.

"Fire the hut!" shouted Woodward, after satisfying himself there was nothing valuable within. "Fire the hut, boys; now we are in possession let us make clean work"—and he seized a brand.

Crone Protests

"Hold! Hellhounds," exclaimed the Crone, in accents deep and hoarse, at the same time, tied as she was, leaping to her feet and standing erect in the center of the apartment; "Eternal vengeance light on him that dare apply the torch! Out upon ye, slaves of fear! Away to your homes! The Crone of the Quequechan neither sues for favor nor regards the hate of man!"

While giving vent to this burst of indignant feeling, every muscle in the body of the Crone expanded: her eyes glowing like balls of fire, her hair, beneath her turban, stood erect, and as she finished, her voice assumed a fullness and depth wholly unlike what had before been heard.

Box Discovered

In the meantime Davis, while rummaging about the bed of straw, discovered a small box which he drew out from what was intended for a pillow. At the sight of this, the Crone bounded forward, seized the box and fell prostrate at the feet of Stephen. In this effort, the cords which confined her hands and feet gave way and let those useful members of human action at full and perfect liberty. Though finding herself thus relieved, she made no effort to escape, but deliberately seated herself upon the rickety stool, which still remained undisturbed in the center of the room, and took from the box a small package, which she handed without uttering a word to Lot Lee.

Note Gives Identity

He opened and read as follows:

"Boston, June ye 10, 1700

"Mary......
"I am in the iron grasp of the king's bloodhounds! Take care of thyself.
KID."

Every eye was instantly turned on the old crone, who still sat on the stool, intent on the effect which the reading of this note might produce.

"And you are ..." said Lee,

"The last mistress of Kid, the pirate!" shouted the hag, snatching the note from the hands of Lee and walking deliberately out of the hut!

No effort was made to detain her; each individual stood riveted to the spot, as if chained by a spell more potent than human will. She mounted the riding ground on the south of her cabin, paused a moment, as if to take last a look in her strange habitation, then dashed off into the high road leading to Newport. This was the last ever heard or seen of the Crone of the Quequechan!

4 King Philip's War

Increasing English settlements along the New England coast in the mid-1600s—including those along the southern boundary of Plymouth Colony at Rehoboth, Swansea ,and Dartmouth—began to lead to friction among the native tribes and the English.

For years after the death of Massasoit in 1661, rumors of war between the Indians and the English became commonplace. Wamsutta—Massasoit's eldest son, his heir, and the second husband of Weetamoe, squaw sachem of the Pocasset—died under mysterious circumstances following a visit to the English authorities. His younger brother, Metacomet or Philip, succeeded him as sagamore of the Wampanoag. Many events, large and small, led to increased tensions between the Wampanoag and the colonists, but it was the events in Swansea in June 1675 that began King Philip's War.

On June 20, 1675, several Pokanoket warriors left their camp at present-day Warren, crossed the Kickamuit River, looted several English homes on Gardner's Neck in Swansea, burned two dwellings, and forced the frightened residents to flee. Three days following the raid, a young man, John Salisbury, shot and wounded a marauding Pokanoket, who subsequently died from his wounds. The next day, Pokanoket warriors attacked and killed seven settlers on Gardner's Neck.

On that same day, according to Phillips in his *History of Fall River,* an English settler named Lawton from Dartmouth was killed in Fall River on the banks of the Quequechan River.

Governor Winslow in Marshfield immediately sent orders to Bridgewater and Taunton to raise a 200-member militia to protect Swansea. However, as the poorly-prepared English militia searched in vain for Philip and his Pokanoket warriors on the peninsula that is now Bristol and Warren, Philip and his men slipped across Mount Hope Bay and landed in present-day Fall River and Tiverton.

There he enlisted the support of his ally Weetamoe, the squaw sachem of the Pocasset. After some wavering and influenced by her younger warriors, Weetamoe joined Philip. Weetamoe had been cold to the English, who she suspected of poisoning her second husband, Wamsutta or Alexander, son of Massasoit and elder brother to King Philip. Her fourth husband, Peter Nunnuit (Petonowett), joined the English, reflecting the split loyalties of the Pocasset and other Wampanoag.

The English militia removed to Taunton, where Plymouth forces under the command of Major James Cudworth joined them. From Taunton, the army marched along the Taunton River to "the great swamp at Pocasset," now known as the Pocasset Cedar Swamp, situated at the Fall River/Tiverton border. A portion of the Pocasset Cedar Swamp is within the watershed of the Quequechan River. On July 19, 1675, the English entered the thick bramble of the swamp and engaged the forces of Philip and Weetamoe. Five member of the English militia were killed and seven wounded, some mortally. All day, the English advanced and the Wampanoag fought a spirited rear-guard action. When night fell, the English retreated from the swamp and the Wampanoag disappeared.

The next day, from his garrison on Mount Hope, Cudworth wrote to Governor Winslow:

> On Monday [July 19th] following we went to see if we could discover Philip, the Bay forces being now with us; and in our march two miles before we came to the place of rendezvous, the captain of the Forlorn was shot down dead; three more were killed or died that night, and five or six more dangerously wounded. The place we found was a hideous dismal swamp [Pocasset Cedar Swamp]; the house or shelter, they had to lodge in, contained, in space, the quantity of four acres of ground, standing thick together; but all women and children fled, only one old man, that we took there, who said that, Wittoma [Weetamoe] was there that day, and that Philip had been there the day before, and that Philip's place of residence was about half a mile off; which we could made no discovery of, because the day was spent, and we having dead men and wounded men to draw off. (70-122)

For a second time, the colonists had let Philip escape. From the Pocasset Swamp, Philip and Weetamoe swept east of the "Watuppa Lake," along what is now Blossom Road and along the Mowry Path, until they reached the Taunton River at the Pocasset ferry landing immediately north of Mother's Brook (later known as Winslow Point at Peace Haven). There

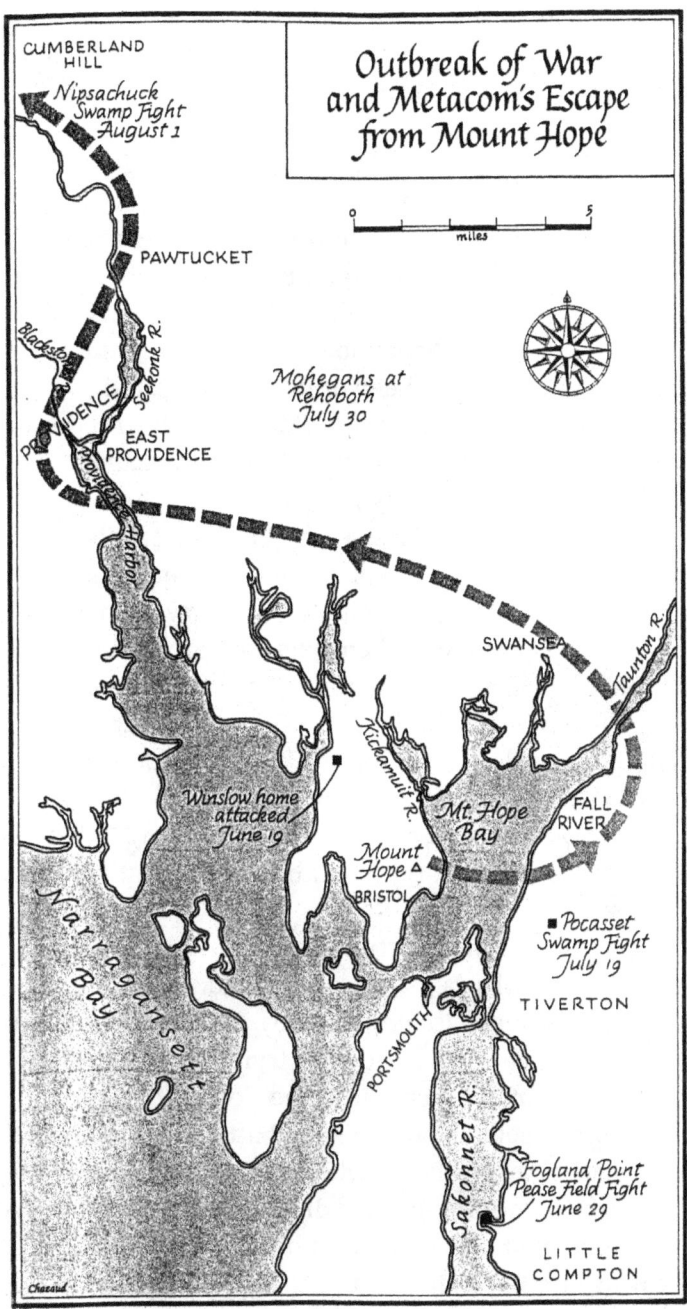

fig. 4.1 **The early battles of King Philip's War and Philip's escape route through Fall River**

Reproduced from *King Philip's War: The History and Legacy of America's Forgotten Conflict*, 1999, by Eric B. Schultz and Michael J. Tougias. This map and related text reprinted with permission of the publisher, The Countryman Press/W.W. Norton & Company, Inc.

they crossed the river and proceeded north into Central Massachusetts where they joined with their Nipmuc allies. The Pocasset warriors who joined Philip were said to number about 300. The map on the previous page shows the probable route of escape of Philip and Weetamoe.

The Pocasset Swamp fight was a pivotal one for the course of the war because it meant that, under Philip's leadership and with Weetamoe's warriors, the combined Pokanoket and Pocasset tribes would expand the war north into all of New England. Any chance of a quick capture of Philip was ended. (70-124) The escape from Winslow Point was singularly fateful for the English and the natives, for it made King Philip's War inevitable. Had the English intercepted and captured Philip and Weetamoe before their crossing of the Taunton River, the history of New England would have been dramatically altered.

The war caused considerable loss of life and widespread destruction of the early English settlements. All of the settlements west of Concord were evacuated.

On February 10, 1676, Philip and his allies set upon Lancaster and destroyed the town. One of the hostages taken was Mrs. Mary Rowlandson, the minister's wife. Mrs. Rowlandson, who was given as the lady's maid to Queen Weetamoe, later wrote an account of her captivity and forced march with Philip's war party. In one of her accounts, she describes Weetamoe as a proud squaw sachem who took particular care of her appearance:

"A severe and proud dame she was, bestowing every day in dressing herself neat as much time as any of the gentry of the land, powdering her hair and painting her face, going with necklaces, with jewels in her ears, and bracelets upon her hands. When she had dressed herself, her work was to make girdles of wampum and beads." (44-45)

The Colonists hated Weetamoe as much as Philip. Cotton Mather said "She was next unto Philip in respect to the mischief that hath been done and the blood that hath been shed in this warr." (44-45)

After a year of successes, Philip's fortunes began reversing. The Colonists and their Indian allies, such as the Mohawks, began to gain ground.

In August of 1676, Weetamoe, accompanied by 26 warriors, took refuge in a dense swamp near Taunton. It appears that she was trying to reach Narragansett territory. However, militia from Taunton, who captured the Pocasset, surprised them. Under the protection of her warriors, Weetamoe escaped and attempted to cross the Taunton River on a makeshift raft to reach Winslow Point, where she and King Philip originally crossed to start the war. She never made the crossing and drowned on August 6th in the river. She was found dead the next morning on the Somerset shore. (70-68)

In *King Philip's War*, Eric B. Schultz and Michael J. Tougias relate the sad end to the proud life of Weetamoe, the squaw sachem:

> Weetamoe's corpse was mutilated and her decapitated head sent to Taunton, where it was placed on a pole, paraded through the streets of Taunton, and left on public display at the Taunton Green. Increase Mather describes the scene: 'The Indians who were prisoners there knew it presently, and made a most

horrid and diabolical lamentation, crying out that it was their Queen's head.'
(70-129)

The death of Weetamoe signaled the end of the Pocasset who followed her and Metacomet; those who were not killed were sold into slavery to the West Indies sugar plantations. Philips' remaining Pokanoket lands in Bristol and Weetamoe's lands in the Pocasset territory in and around Fall River were confiscated by the Colonists and sold to pay war debts. Those Pocasset who sided with the English were settled in 1704 in the area now known as the Maplewood section of Fall River along Stafford Road.

However, the lingering bitterness from the atrocities committed by both sides during King Philip's War caused constant friction between the Indians and the Colonists. As a consequence, the Indians asked to be moved to a more distant location. Benjamin Church facilitated their relocation to the eastern side of North Watuppa Pond on land that he owned, what later became known as the Watuppa Reservation and which is today part of the watershed of the City of Fall River's water supply.

Philip was hunted down and killed in a wooded swamp in Pokanoket territory on August 12, 1676, by a Pocasset warrior, John Alderman, who had remained loyal to the English. Colonel Benjamin Church of Little Compton pursued Philip and found him in what is now Bristol.

Philip's body was drawn and quartered and a piece hung on each of four trees. His head was decapitated and his skull placed on public display on a pole in Plymouth for a whole generation—25 years. His wife and nine-year-old son were sold into slavery. When Cotton Mather heard of the news of Philip's death, he exclaimed from his pulpit "God has sent us the head of Leviathan for a feast." (44-55)

Benjamin Church later become active in the commercial affairs of the village that was to become Fall River. About 1680, he built a home for his family at what is now the corner of Pond and Anawan Streets, near the lower falls of the Quequechan River and, with his brother Caleb, built and operated a saw mill, a grist mill, and a fulling mill on the upper falls of the river. Colonel Church, his wife, and children are buried in the churchyard at Little Compton Common.

Massasoit's generosity in selling so much of his domain to the English eventually physically constrained Philip and the Wampanoag. The idea of land ownership was foreign to the native tribes, who at first thought that they were sharing their land with the English. As more land was "sold" to the English, a final conflict was inevitable.

Before the war, about 1670, when Metacomet had received yet another complaint from the English council at Plymouth, Philip very simply and eloquently relayed to his friend John Borden of Portsmouth the complaints and grievances of the Indians:

> The English who came first to this country were but a handful of people, forlorn, poor and distressed. My father was their sachem. He relieved their distresses in the most kind and hospitable manner. He gave them land to build and plant upon. He did all in his power to serve them. Others of their own countrymen

came and joined them. Their numbers rapidly increased. My father's counselors became uneasy and alarmed, lest, as they were possessed of firearms, which was not the case with the Indians, and take from them their country. They, therefore, advised him to destroy them, before they should become too strong, and it should be too late.

My father was also the father of the English. He represented to his counselors and warriors that the English knew many sciences which the Indians did not; that they improved and cultivated the earth, and raised cattle and fruits, and that there was sufficient room in the country for both the English and the Indians. His advice prevailed. It was concluded to give victuals to the English. They flourished and increased. Experience taught that the advice of my father's counselors was right.

By various means they got possessed of a great part of his territory. But he still remained their friend til he died. My elder brother became Sachem. They pretended to suspect him of evil designs against them. He was seized and confined, and thereby I became Sachem. They disarmed all my people. They tried my people by their own laws, and assessed damages against them which they could not pay. Their land was taken. At length a line of division was agreed upon between the English and my people, and I myself was to be responsible. Sometimes the cattle of the English would come into the cornfields of my people, for they did not make fences like the English.

I must then be seized and confined till I had sold another tract of my country for satisfaction of all damages and costs.

Thus tract after tract is gone. But a small part of the dominion of my ancestors remains.

I am determined not to live till I have no country. (31-2)

5 The Colonial Era in Freetown

The Town's Early Economy

The first permanent settlers to the area that became Freetown arrived around 1680. There appears to be two sources of immigration: one from the Plymouth Colony (the Pilgrim Congregationalists) and one from Newport (the Quakers).

Puritan immigration to New England was founded on religious impulses, with the goal of establishing model Christian communities that would allow the emergence of a new individual, untainted by the corruption of European society. Therefore, before the General Court would approve a new settlement, the environment of the area to be settled had to sustain approximately 30 to 40 families, or enough to support a minister. The Puritan magistrates were determined that none of its model communities be without the religious oversight of a spiritual leader.

There were two environmental requirements for new settlements: (1) they should contain adequate salt marsh and (2) suitable upland grazing land. The upland would provide cropland and pasture land for livestock during the growing season, while the marshes would provide the salt hay for winter fodder. Unlike the plantation economies of the southern American colonies and the West Indies, the Puritans (and Pilgrims) decided that the most appropriate economic base for their New England would be fisheries and farming, particularly the growing of livestock. Locations along the coast that had extensive salt marshes, such as Rehoboth, Swansea, and Dartmouth, were especially attractive and developed early communities in the 1660s. Later, the salt marshes of Assonet Bay and its upland fields became recognized as a prime site for a new community.

Coastal locations were also the easiest to access. Before the emergence of the railroads in the 1840s, most population centers in the Colonial Era grew along navigable waterways, since water vessels provided the fastest, easiest, and least costly way of moving goods between cities and regions. Roadways were too rudimentary to carry major amounts of freight over long distances.

As with most New England towns, the main economy of Freetown was subsistence agriculture. Virtually all of a family's needs were met through its own resourcefulness. Farm families grew their own food; raised their own livestock for meat, milk, butter, cheese, and eggs; made their own clothing on foot-powered spinning wheels and looms from wool provided by their own sheep; grew apples for their hard cider; made their homes, barns ,and tools from lumber provided by their own wood lots; obtained fuel from those same wood lots; and generally provided for themselves.

A Scotsman, Chancellor Livingston, observed in 1813 that the Yankee farmer "can mend his plough, erect his walls, thrash his corn, handle his axe, his hoe, his sithe, his saw, break a colt, or drive a team, with equal address; being habituated from early life to rely on himself he acquires a skill in every branch of his profession, which is unknown in countries where labor is more divided."

In order to raise cash for those items and services that had to be purchased by the farmer, he engaged in cottage industries such as making shoes, hats, or other items. He also sold his surplus agricultural goods for this purpose. The items and services that had to be purchased included iron parts of tools from the local blacksmith, the grinding of corn and grains at the local grist mill, the sawing of lumber at the saw mill, the finishing of their home-made cloth at the fulling mill, and other goods such as salt, sugar and rum.

The early industries in the three villages in Freetown on the Taunton River—Assonet, Steep Brook, and Fall River—were all related to agriculture and were the same as in all other New England communities: a saw mill made lumber from locally-cut logs, a grist mill ground the farmer's grains and corn; a fulling mill processed hand woven woolen cloth into a finer, more durable and workable fabric; and a blacksmith shop was the iron works that provided the iron parts for tools, plows, harnesses, and household implements, in addition to horseshoes.

These early industries required waterpower to move the water wheels that powered the saws and grinding wheels of these early mills. The Taunton River was too large and with too little gradient to harness for waterpower and, even if it could, the resulting flooding would cause the loss of thousands of acres of land. However, the tributaries of the Taunton

River were ideal for this purpose. They had sufficient gradient or "fall," were small enough to be harnessed with small dams, and the amount of water flowing through the streams was manageable for these early mills. The tributaries of the Taunton River in Freetown that became important for mills were the Assonet River, Rattlesnake Brook, Steep Brook, and the Quequechan River.

The specialized plantation economies of the West Indian sugar islands and the southern American colonies (cotton and rice) could not supply themselves with sufficient food and other basic items. Resourceful New Englanders saw a market and responded. As a result, New England became the breadbasket of the plantation economies. New England farmers provided these southern monocultures with essentials such as flour, biscuits, salted beef, salted pork, salted fish, butter, cheese, peas, rooted vegetables, lumber, and barrel staves, as well as horses and other livestock.

In 1689, in *Brief Relation of New England*, a contemporary remarks that:

> The other American plantations cannot well subsist without New England, which is by a thousand leagues nearer to them than either England or Ireland; so that they are supplied with provisions, beef, pork, meal, fish, &c, also with the lumber trade, deal boards, pipe [barrel] staves, &c, chiefly from New England. Also the Caribee Islands have their horses from thence. It is then, in a great part, by means of New England, that the other plantations are made prosperous and beneficial. (9-42)

fig. 5.1 **Typical American schooner in the Colonial Era**

The American schooner *Baltick*, coming out of St. Eustatia in 1765, was a typical schooner in the coastal and West India trades.

In 1660, Governor Winthrop mentioned that the Massachusetts Bay Colony had prospered to such an extent that it was now able to export farm products and build a shipping industry, to the extent that "the country doth send out great store of biscott, flower, peas, beife, porke, butter, & other provisions to the supply of Barbados, Newfoundland, & other places, besides the furnishing out many vessels & fishing boats of their owne, so as those who come over may supply themselves at very reasonable rates." (9-43) Resourceful New Englanders supplied provisions not only to the West Indian Islands and southern colonies but also north to the fishing villages of Newfoundland.

The trading between New England and the West Indies and the southern colonies in farm products and fish (the famous salt cod) required large numbers of marine vessels of all kinds, sparking the growth of the New England shipbuilding enterprises in the 1600s, including shipyards all along the length of the main stem of the Taunton River, from Bridgewater to Mount Hope Bay, but particularly in Somerset, Dighton, Assonet, and Steep Brook. This shipbuilding activity expanded into larger vessels as trade with Europe, Africa and the West Indies increased. By the 1700s, Somerset, Assonet, and Dighton had become major shipbuilding centers on the Taunton River, with minor shipbuilding occurring at Steep Brook and Fall River.

Shipbuilding began early on the Taunton River, probably about 1680. At first, small vessels were built to serve local needs, such as fishing. Later, larger vessels were built to transport firewood, building lumber, and farm produce to the emerging towns of Providence, Bristol and Newport. In the 1700s, this shipbuilding activity on the Taunton River grew to serve the coastal trade, serving cities such as New York, Philadelphia, Baltimore and Savannah.

Ships involved in the coastal shipping trade proceeded up a river to its most navigable point, in order to reach markets as far inland as possible. During the Colonial period, the farthest navigable point on the Taunton River was in Dighton, a few miles above where the Assonet River enters the Taunton. With its inland location and its protected harbor, Assonet therefore became the transfer point for goods brought by oxen teams from inland sites such as Taunton, Middleborough, Berkley, and even Fall River. Goods such as farm produce, manufactured products and wood were loaded at Assonet for shipment to points south, including Providence, Bristol, Newport, New York, Philadelphia, Baltimore, Charleston, Savannah, the West Indies, and foreign ports.

In return, ships brought back salt, molasses, sugar, tobacco, rum, tropical fruits, rice, and coffee from the West Indies, the southern American colonies and even South America.

Vessels from the Taunton River seemed to have a particular affinity for the southern colonies, particularly in Georgia, where they spent the winter and then returned in the spring to be outfitted in Assonet and Somerset. A commentator in 1897 remarked that "to Georgia were shipped beef, pork, livestock, dairy products, apples, hay, and in fact much the same products that went to the West Indies. The rice, indigo and cotton plantations of the southern coastal plains were comparable in many respects to the sugar plantations. They found specialization in their 'cash' crops so profitable that it seemed as if they could not

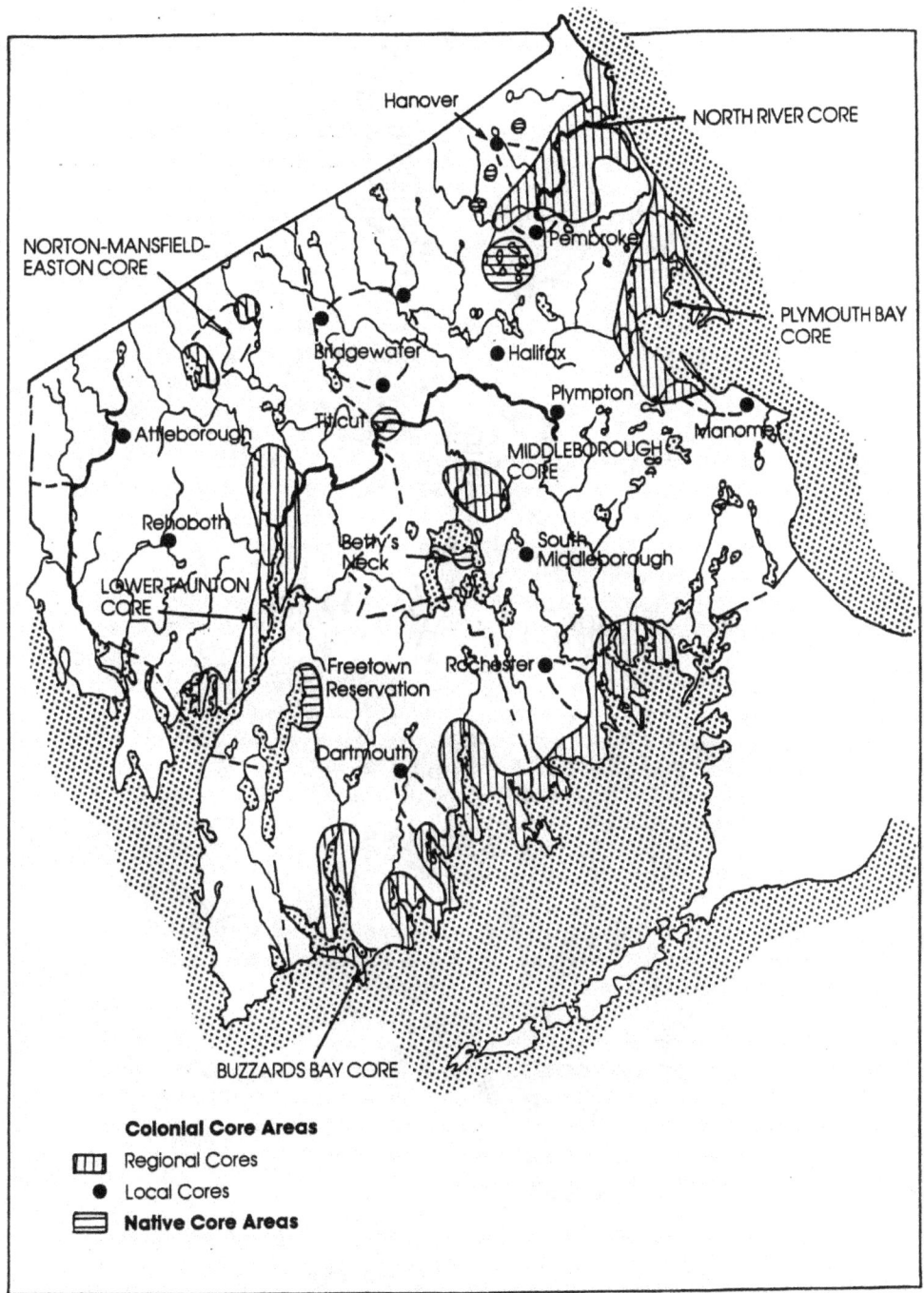

fig. 5.2 **Colonial period core areas**

Source: Massachusetts Historical Commission: *Historic and Archaeological Resources of Southeast Massachusetts.*

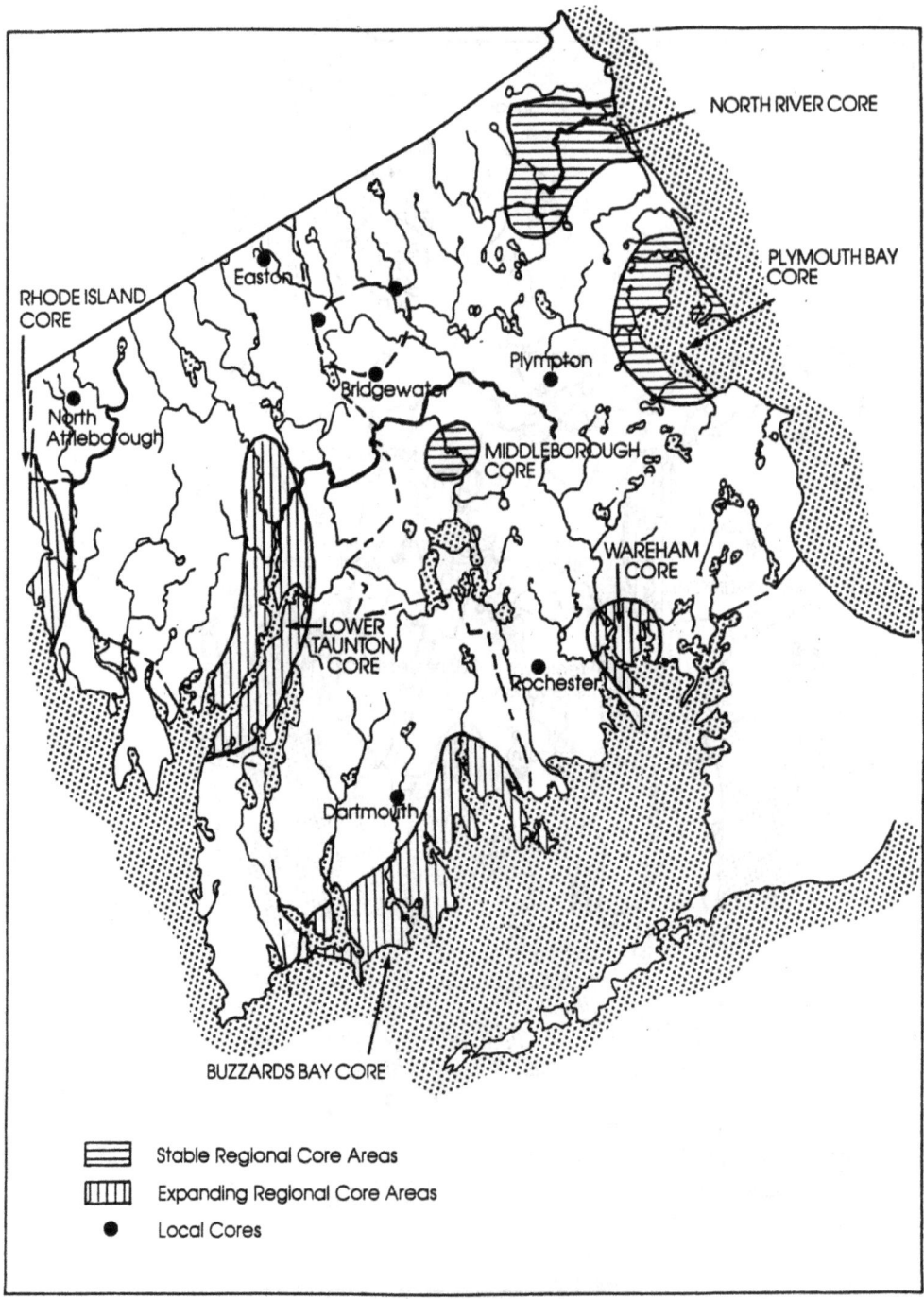

fig. 5.3 **Federal period core areas**

Source: Massachusetts Historical Commission: *Historic and Archaeological Resources of Southeast Massachusetts.*

afford to raise their own food supplies." (9-143)

Coastal trading vessels built in Assonet and Somerset typically had drafts of only four feet, in order to navigate the shallow waters of the southern inland waterways where the plantations were located.

In his *History of the Town of Somerset*, William A. Hart states that the majority of the New England shipping fleet would winter in the south and that "Darien, Georgia was the favorite wintering spot for vessels from Mt. Hope waters since they could always find there the familiar craft and faces of their home neighborhood. For many years, it was said that if during the winter you wanted news of Assonet, Freetown, Dighton or Somerset you would do best to write to Georgia for it." (41-74)

By 1760, the ports of Dighton, Somerset, and Assonet equaled Plymouth in the number of vessels and goods transported. The maps on the preceding pages show the growing importance of the Taunton River corridor in the region during the Colonial and Federal periods.

Steep Brook in the Colonial Era

Steep Brook was one of the villages in Freetown that had a history similar to Assonet and which harbored an active rivalry with Assonet. Steep Brook was also a location for shipbuilding, but lacked Assonet's protected harbor.

Because of its location, Steep Brook became the hub of travel on land and on the Taunton River. As early as 1808, stagecoaches began regular service between Boston and Newport via Fall River, first on alternating days then on a daily basis. The stagecoaches stopped at the Green Dragon Inn located at the corner of North Main Street and Wilson Road to change horses, to check equipment at the blacksmith shop, and for passengers to rest and dine. The sign on the inn said "Beer, Oysters and Horsekeeping." From Newport, connections could be made with New York on fast sailing packets. Sailing from Newport saved two or three days on the Boston to New York route over the grueling shore stage line, depending on the season and road conditions. (65-29)

Previous to 1808, the Boston to Newport stagecoach operated via Attleborough, East Providence and Bristol Ferry, where a somewhat undependable ferry brought passengers over to another waiting stagecoach on Aquidneck Island. However, with the building of a bridge at Tiverton Narrows to the island (the original Old Stone Bridge), the route changed and came through Taunton and Fall River. (65-26)

The stage left Worthington's Tavern on Bromfield Street in Boston at 5:00 am every day except Sundays and arrived at Newport at 6:00 pm, if not delayed. The stages carried six passengers and were often followed by a baggage wagon. (65-26)

Once in Newport, passengers boarded a packet to New York. In its dependability and comfort, the stage line to Newport and the fast packet to New York were the predecessors of the "Boat Train" and the Fall River Line to New York.

At the bottom of Wilson Road on the Taunton River, Chace's Ferry carried passengers across the Taunton River from Dighton and Somerset where connections could be made with the stage for either Boston or Newport.

In addition, Steep Brook was a connection for stagecoaches running from New Bedford and Providence. A bridge crossing at the Narrows was not made until 1828, so the only practical way for freight and passenger travel in an east/west direction from New Bedford to Fall River and Providence was via Wilson Road at Steep Brook. In 1825, regular stagecoach service began between New Bedford and Providence, with passengers crossing at Slade's Ferry (site of the old Brightman Street Bridge).

At first, passengers crossed the Taunton River by small sailboats and by changing stagecoaches, since early sailing ferry craft were not stable enough to safely carry carriages and horses. Providence coaches unloaded passengers on the Somerset side and New Bedford coaches unloaded on the Fall River side, with the coaches returning to their respective destinations.

In order to have continuous stagecoach travel between New Bedford and Providence, sailboat ferries were replaced in the next year, 1826, with larger horse-powered ferries that allowed horses and coaches to cross the Taunton River at Slades Ferry. (41-128)

Steep Brook was the site of much shipbuilding activity. At Miller's Cove, where the Fall River Country Club is now located, there existed from the early 1700s Thurston's Wharf, where vessels were made for the West Indies trade. At this location, there was also a cooper's shop, where barrels were made to transport goods to and from the West Indies and southern colonies. There was also a tavern located at this location until 1776, when it was destroyed by fire.

Next to Thurston's Wharf was Miller's Wharf, on the cove that bears his name. Robert Miller was a shipwright and built a shipyard here after purchasing the lot in 1738. Robert's son and grandson enlarged the shipyard in subsequent years. Arthur Phillips, in his *History of Fall River*, states that many vessels were built here, including the "George Washington," built in 1838. That vessel made trips down the Taunton River to Charleston, Savannah, and Cuba to trade in cotton and rice. In 1849, in the gold rush of '49, Captain Miller sent the George Washington around Cape Horn to the California gold fields with a crew from Steep Brook. (63-92)

Fall River in the Colonial Era

During the Colonial Era, the location of the "West End" of Freetown above and below the Quequechan falls on the Taunton River did not lend itself to development. The village lacked an adequate protected harbor and was subject to westerly and southwesterly winds from Mount Hope Bay. The steep slopes rising out of the Taunton River—a unique situation along the otherwise flat landscape of the river—put this location at a disadvantage because it meant that to access the interior required "double teaming" of oxen up the steep hill. It was also not as inland as the villages of Steep Brook and especially Assonet, which was in close proximity to large areas of fertile agricultural land. The wilderness above the Quequechan falls had rocky till soils that were difficult to farm.

Like all rivers, the Quequechan River is subject to seasonal flows, with heavy flows during the spring and wet seasons and low flow during the dry summer season or during drought. About 1700, a dam was built on the west side of North Main Street to provide a more even flow for the two saw mills, two grist mills, and a fulling mill that were located on the falls. This dam raised the level of the water above the falls two feet above its natural elevation. When the Troy Cotton and Woolen Manufactory was built about 1813, the dam was replaced with a new one that raised the river's water level above the falls to three feet above the original level of flow.

There were two taverns in Fall River, one at Brightman Street near the Slade's Ferry operated by the Brightman family, and one at the corner of North Main and Central Streets operated in 1738 by Stephen Borden. The Borden tavern was a popular place until it was replaced in favor in 1803 or 1806 by the "Mansion House" on Central Street, around the corner from North Main Street, operated by Phoebe Borden. In 1807, she married Major Bradford Durfee and they operated an inn there until 1828. (65-29)

A small water wheel or "wash wheel" was located on the upper Quequechan falls near Main Street. The wash wheel raised water from the stream below and brought it up for housewives to use for laundry purposes. (63-76)

Near the river was the Methodist Meeting House. The Methodist minister, Father Taylor, presided over many a baptism by immersion in the Quequechan.

In 1813, as the little village was about to enter the dawn of the Industrial Revolution, there were 30 dwellings surrounding the river with about 300 inhabitants and a school house, three saw mills, four grist mills, one fulling mill, three blacksmith shops, and several small stores. (25-10)

At this time, the only shipping from the village was a few small sloops that carried cordwood to Newport and Bristol and the local sloops that shipped corn to the Quequechan River gristmills and sent back flour to the Rhode Island towns. Because of the lack of wharves on the Taunton River at the village, cotton for the early mills and other merchandise shipped in and out of the village had to be loaded at the wharf two miles upriver at Slades Ferry at Brightman Street or at Steep Brook. This merchandise arrived on ships from Providence that plied the Taunton River between Providence and Taunton.

Direct shipping between Fall River village and Providence began as the village expanded and the first textile mills on the Quequechan created a steady demand for cotton. The first direct service to Providence was on a small schooner large enough to hold 10 bales of cotton and a small additional cargo of flour and miscellaneous goods. Ever-larger sailing vessels succeeded this schooner until the steamer Hancock was put into service in 1828. (25-10)

Alfred J. Lima

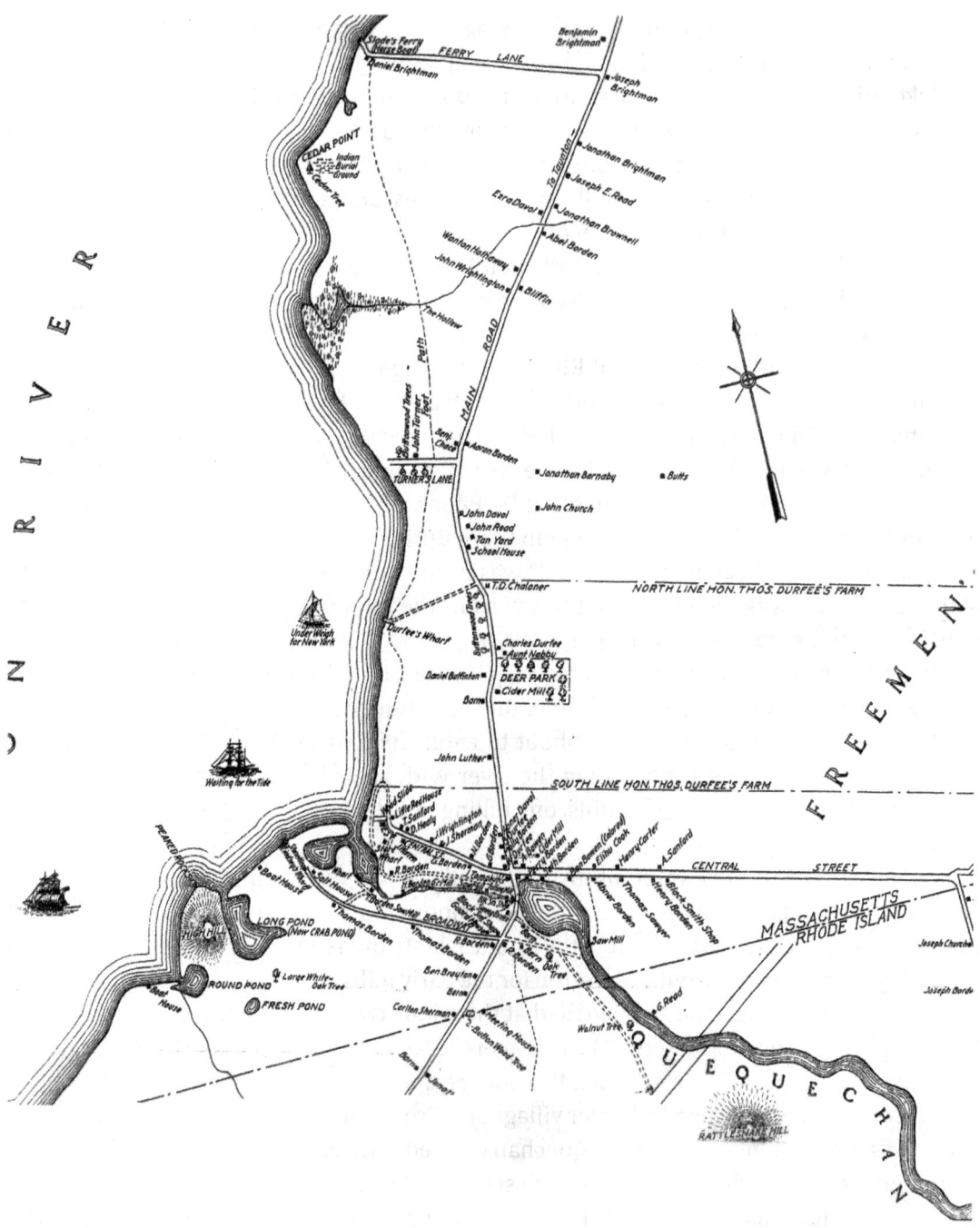

fig. 5.4 **The West End in 1812**

This is the village as it appeared at the close of the Colonial era. The next year, the first textile mill was to be built in the village, ushering in the advent of the Industrial Revolution to Fall River.

6 Colonial Industries on the Quequechan River

The Saw Mills

The Quequechan River provided power for the usual Colonial industries of saw mill, grist mill, and fulling mill. About 1680, following King Philip's War, Benjamin Church established, with his brother Caleb from Watertown, a saw and grist mill on the south side of the river where it enters the Taunton River. These mills were later purchased and operated by Thomas and Joseph Borden. Joseph Borden also operated a fulling mill at one of the falls near the top of the Quequechan. Steven Borden owned the north side of the Quequechan River and operated a saw mill and a grist mill there. (25-25)

The profits developed by the Bordens, Durfees, and Bowens from these early enterprises grew over time into a sizable amount of the capital that later became invested in textile and iron manufacturing. The technical and entrepreneurial skills learned by these capitalists in these early businesses became important when the Industrial Revolution came to the town. There were few specialists in the early years; everyone learned to do almost everything.

Even during the early years of development on the Quequechan River, the need for

fig. 6.1 **Sawyers sawing a log in a saw pit**

Source: Tunis: *Colonial Craftsmen and the Beginnings of American Industry*

building materials for local use and for export to the growing towns of Newport and Bristol required prodigious amounts of building materials, especially lumber.

The preparation of logs into sawn lumber went through various technological evolutions. Initially, logs were sawn in "saw pits." These pits were holes dug into the ground with logs placed over them. A man in the pit and one above worked the two-man saw.

These cutters were called "sawyers" and were a recognized craft. The leader of the team, the top sawyer, stood on the timber and guided the saw along a snapped-on line. He also pulled the saw up so that the pitman below could pull it down again for the cutting stoke. Rollers allowed the timber to move along the pit. (79-26) It was a dirty and dangerous occupation. There were very likely one or more such pits in the area during the early years of settlement, but their locations have not been recorded.

With the development of the up-and-down saw mill, using multiple gang saws, the Quequechan River was harnessed for powering saw mills. Gang saws operated by water power began appearing in Maine as early as 1650. The figure on the next page shows the workings of such a mill. These up-and-down saw mills were slow by modern standards, for they cut only 25 to 50 board feet a day. At that speed, it would take a year to cut enough wood to build one house. (62a) An operating up-and-down saw mill still operates at the Taylor Mill site in Ballard State Forest in Derry, New Hampshire. There was no further improvement in cutting lumber until the circular saw was introduced 164 years later in 1814.

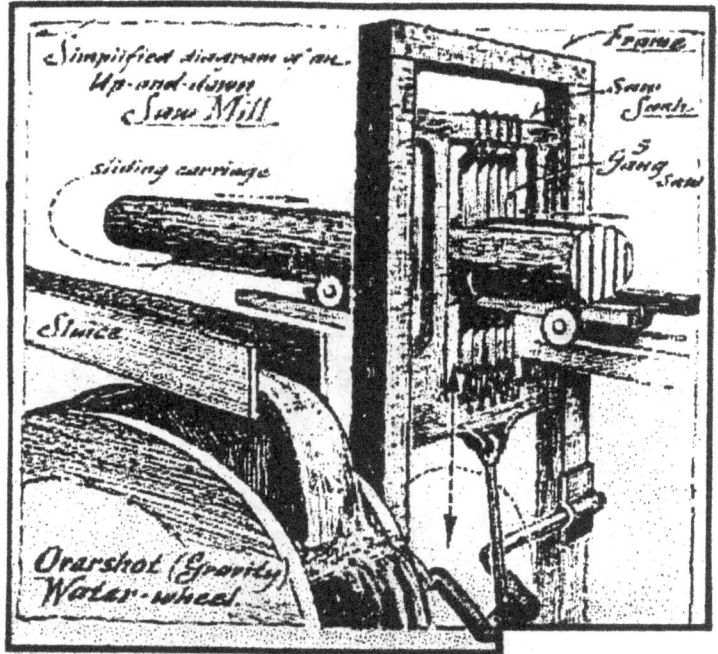

fig. 6.2 **Workings of a typical gang saw powered by water on the Quequechan River**

The Grist Mills

In his *Centennial History of Fall River, Massachusetts*, Henry H. Earl describes the early industry of the young Richard Borden, who at 17-years-of-age began operating (from 1812 to 1820) a grist mill on the Quequechan River with his brother Jefferson Borden.

The Borden brothers would:

> go down to Prudence and Conanicut [Jamestown] Islands, in the sloop Irene and Betsy, which carried about 250 bushels of corn, and having secured a load, to return to Fall River and tie up at a little wharf within the creek, and discharge directly into the mill. The Irene and Betsy was also a sort of packet between Fall River and the neighboring places, and the surplus meal was sold in Warren, Bristol or Providence, and a return freight secured, of provisions, groceries, cotton, etc. (25-47)

As business improved, another grist mill was built at the next-highest falls on the Quequechan. In order to facilitate the movement of corn from the sloop on the Taunton River to the mill on the Quequechan, a tram was built with a car running up and down the hill, pulled by a rope and operated by the same water wheel that powered the grist mill. (25-47)

fig. 6.3 **The workings of a typical New England grist mill**

The hopper platform would normally stand on the "stone-floor" inside the mill instead of floating in the air over the millrace, as it does here. It is put there for compactness and clarity.

Source: Tunis: *Colonial Craftsmen and the Beginnings of American Industry*

It is interesting to note the distinction between the grist mills built on the Quequechan River and the mills on other streams. Usually, grist mills served the farms that surrounded them. However, farming was difficult on the thin soils that overlay the granite ledge that surrounded the river. The Quequechan River produced more power than could be used by the meager West End farms. The islands on Narragansett Bay, however, had the opposite problem: lots of productive farmland but few streams capable of operating a grist mill. The solution by the enterprising Bordens was to bring the islands' grain to the Quequechan River, grind it into meal, and then market it for the farmers at local towns. They therefore became not just millers for the whole region but merchants as well.

In 1813, there were four grist mills in the village of Fall River, more than would be expected for a small village. (25-10)

Operators of grist mills were called millwrights. However, typically the millwright was also a carpenter who specialized in setting up mills and water wheels and who had a working knowledge of gear ratios, drive-shaft speeds, and other equations. The experience that Richard and Jefferson Borden gained as millwrights provided the expertise for developing early water-powered textile mills in the town. (25-47)

The Fulling Mills

Once woolen cloth was woven at home, it was taken to the fulling mill on the Quequechan River operated by Joseph Borden on the south side of the river at the top of the falls near Main Street. In his *History of Fall River*, Phillips states that nearly all of the woolen cloth woven in Fall River was cleansed and fulled at this mill. The mill was destroyed in the great fire of 1843. In the village of Steep Brook, a fulling mill was also operated by Barnabas Clark, which was located on the Steep Brook stream, just below where Highland Avenue is now located.

Fulling the cloth began with a thorough washing with hot water and soap to remove dirt and some of the grease. Then it went into the beating trough with fuller's earth and was "thumped" mechanically for hours. Fuller's earth was a greenish absorbent clay that took out almost all of the grease from the wool. The earth then had to be washed out of the cloth. Fulling mills were always located on streams because of the need for process water and

Fulling mill

fig. 6.4 **The mechanism operating thumpers in a fulling mill**

Note the water wheel to the left that powers the thumpers.
Source: Tunis: *Colonial Craftsmen and the Beginnings of American Industry*

for power to operate the "thumpers." The fulling mill cleansed the woolen cloth of grease and lanolin, compacted its fibers, raised the nap of its surface, and generally made the cloth usable and marketable.

The fuller then stretched the wet cloth on tenter frames to remove any wrinkles and to maintain the cloth's width. When the cloth dried, the fuller hung it over rods and curried its surface thoroughly. The final step required the fuller to trim off the shagginess with long shears. (79-35)

fig. 6.5 **Illustration of a fuller at work**

The Blacksmith

The blacksmith was perhaps the most important of Colonial artisans because few other men had the skills to do the work of the blacksmith and because he was essential to so many enterprises. Every village needed a blacksmith and many communities gave land grants to attract them. (74-14) The map of Fall River in 1812 shows that there were two blacksmith shops on the upper falls of the Quequechan River and one other in the village.

Water power would have been important to those blacksmiths who needed to refine pig iron into malleable wrought iron in their own forges. For these blacksmiths, water wheels powered bellows, trip hammers, and grinding wheels, the latter used to grind and polish finished iron products.

While a substantial part of the blacksmith's work was in shoeing horses and oxen, the iron products he produced were essential to all other Colonial occupations, including the farmer, shipbuilder, and housewright, among others. He made the iron parts of farm implements, such as axes and plows, and the iron essentials for home construction, including nails, latches, hinges, gates, fences, and andirons. He made iron rims for the wheelwright's wooden wheels and iron hoops for the cooper's wooden barrels. In shipbuilding areas, such as along the Taunton River, the blacksmith was especially invaluable because of his ability to manufacture the essential iron components of vessels, including ship fittings and hardware, anchors and anchor chains, rudder irons, shipwright tools, and hooks, rings, and bolts.

In Colonial times, if a housewife wanted a long fork to use in fireplace cooking, her husband would order it from the local blacksmith. There were no utensils ready-made in the blacksmith shop; all were made to order.

Colonial travel on unpaved roads was challenging for horses, stagecoaches, and baggage wagons, and one of the duties of the blacksmith was to repair equipment at stops along the route of the stagecoach. In his *History of Fall River*, Phillips mentions that when the Boston to Newport stagecoach stopped at the Green Dragon Inn in Steep Brook, at the corner of North Main Street and Wilson Road, passengers rested and dined while the village blacksmith checked the coaches and wagons for needed repairs.

Blacksmiths were the machinists of their day and, together with millwrights, were

fig. 6.6 **The blacksmith shop**

This is a sketch of an actual blacksmith's forge in the cellar of the Old Cook Tavern in Plainville, Connecticut, circa 1790. In the days of stagecoach travel, it was important that repairs and adjustments be made as quickly as possible between stopovers. Charcoal from local forests was used universally as fuel in the blacksmith's forges, since it burned cleaner and hotter than wood. A blacksmith's apprentice operated the bellows.

Source: Sketch reproduced from Sonn: *Early American Wrought Iron*

indispensable in propelling the village of Fall River into the Industrial Revolution. Blacksmiths were responsible for constructing the early textile spinning and weaving equipment that were placed in the town's first mills. As the Industrial Revolution progressed and the design of machinery became more complex, the role of the blacksmith evolved into that of the machinist. However, the blacksmith remained a viable occupation in the city until well into the twentieth century, when the internal combustion engine eventually replaced horsepower.

One of the blacksmiths on the Quequechan River was a "Father Healy," who "had a very large family and one of his descendants was Joseph Healy, who became a prominent citizen and manufacturer." (63-75)

Blacksmithing required a source of iron and, along the New England coast, iron making became an important Colonial industry that provided the blacksmith with his raw product.

Shipbuilding

During the Colonial Era, every farmer of substance along the Taunton River was also a shipwright. He built his own sloop of 35 or 40 tons designed to carry his produce to local destinations such as Providence, Bristol, Warren, Newport, and sometimes even to New York. Some members of the family were seamen and some joined the West Indies trade, or became privateers. Others became members of the new Revolutionary Navy. (31-12)

While Assonet, and to a lesser extent Steep Brook, were Freetown's main shipbuilding centers, shipbuilding also occurred at Fall River. Several wharves lined the shore of the Taunton River north of the mouth of the Quequechan, and a wharf dotted the shoreline at the base of every farm along the river. The long narrow parcels of the Freeman's Purchase that extend from the Taunton River eastward into the interior for miles represent the importance of access to the river for Colonial farmers.

fig. 6.7 **A typical New England shipyard**

Source: Tunis: *Colonial Craftsmen and the Beginnings of American Industry*

As noted earlier, Colonel Richard Borden and Major Bradford Durfee built vessels at the mouth of the Quequechan River, an occupation that led them to forge their own iron parts for their ships and which evolved into the formation of the Fall River Iron Works.

Later, the Fall River Iron Works Company built its own vessels on its site at the base of the Quequechan River to transport raw materials to its factory (coal, iron, iron ore, etc.) and to bring away its finished products to various local and distant markets.

Whaling in Fall River

One of the early local industries of the area was whaling. At first, whaling was a land-based occupation, where whales were taken in Narragansett Bay by small boats. The earliest ocean-going whaling activity in the area was recorded in 1733, when the first regularly equipped whaleman arrived in Newport. By 1750, Bristol and Warren were thriving whaling ports. However, Tiverton, Assonet, Steep Brook, and the West End or Fall River also had whalers based at their ports. The Revolutionary War resulted in the temporary demise of the whaling industry in America and, when the war was over, the whales in Narragansett Bay had gone to other feeding grounds. (61-367)

In 1841, the historian Orin Fowler noted that 90 men and boys were engaged in whaling in five ships based in the Town of Fall River. Another 100 sailed on other ships based in the town and 200 sailed from ships in other ports, for a total of 390 seamen from Fall River engaged in whaling. There were a total of 541 seamen employed in the town. (38-34)

On Pardee's Wharf on the Taunton River north of the mouth of the Quequechan River, there existed a whale oil manufactory, which processed 32,000 gallons of oil annually. (38-33) Dr. Nathan Durfee fitted out several whaling vessels based in the port of Fall River and established an oil works on the waterfront. (25-54)

After 1775, in its heyday, Westport had 20 to 30 whaling ships based in that town. While this whaling activity was dwarfed by New Bedford's whaling industry of over 300 vessels, whaling was nonetheless also a valuable part of the history of Fall River.

Coopering

Packaging of retail goods, as we are familiar with today, did not exist before the invention and widespread use of cardboard and corrugated boxes. Until well into the twentieth century, most goods were delivered in bulk—in wooden barrels—and doled out in small amounts in retail stores. Where retail boxes were used—as in shoes and for some grocery items, such as salted fish—the boxes were wooden.

Mr. Fernand Auclair remembers many items that came in barrels to his father's grocery store, Auclair's Market, at 64 Brightman Street, on the corner of Leonard Street near the Brightman Street Bridge in the city's North End. In the 1920s, 1930s, and even the 1940s, items that came in barrels, tubs, or wooden boxes included dried peas and beans, pickles, lard, butter, tea from China, chewing tobacco, salted fish, molasses, and vinegar. Even fresh

fig. 6.8 **Cooper assembling the staves of a barrel**

Source: Tunis: *Colonial Craftsmen and the Beginnings of American Industry*

de-feathered turkeys came in barrels. Items were either scooped out in small bags for sale or were measured out as requested. (89) Joseph Jean remembers working as a young boy at LaJeunes' Grocery Store in the Flint, doing such jobs as taking sugar from large barrels and filling paper bags for retail sale.

Virtually everything in Colonial times was shipped in barrels, including whale oil, salt, sugar, salted fish, salted beef and pork, root vegetables, fruit, dried produce such as peas and beans, tobacco, rice, wheat, corn, flour, and hardware.

Smaller versions of barrels were called kegs and used for especially heavy items such as nails and screws and for rum and gunpowder. Household and farm tasks also required the smaller version of barrels: wooden tubs and pails. (79-22)

"Slack" barrels were used for dry goods such as flour, corn meal, and threshed grain. "Wet" barrels were used for molasses, maple syrup, cider, beer, salted fish and meat, turpentine, and tar. Wet barrel staves were one-inch thick and made from oak. Slack barrels could be made from softer wood. Hoops were made from strips of hickory or chestnut; only after 1800 did iron hoops begin to be used. (79-22)

Barrel staves and larger pipe staves in the rough were among the earliest exports from the Colonies. Ships were required to bring staves on every shipment to England, which had depleted its forests and needed staves for its exports, and to the West Indies for its exports of sugar and salt.

The work of the cooper required long experience, consistent skill and a sharp eye. Barrel staves had to be finished precisely and beveled to the radii of the barrel. A finished barrel had to be watertight and allow no infiltration of air or pests. (79-22)

New England farmers split and finished oak barrel staves as winter work and supplied the cooper with sufficient staves to pay for the barrels that the cooper made for them. In communities such as Freetown and Fall River, where trade and later industrialization created the demand for many barrels for shipping goods, the importance of the cooper increased over time. The Fall River Iron Works, for example, employed many coopers to produce the kegs needed to ship its nails and other products. (38-33)

Colonial Textile Production

On New England's self-sufficient farms, textile production was a family enterprise. The principal fabrics were woolens, made from wool sheared from the farm's sheep, or linen,

fig. 6.9 **Home textile production in Colonial America**

Hand carding (left) aligns the fibers, which are then spun on a wool wheel (center) and woven on the loom (right). Colonists used flax and wool to make their clothing, for cotton was not available on Northern subsistence farms and was, in addition, not suited to the methods used in home cloth production.

Source: Merrimack Valley Textile Museum, as shown in *Run of the Mill*

produced from flax grown on the farm. In 1839, 1,138 pounds of wool were sheared from sheep in Fall River.

After shearing his sheep and washing the wool, the farmer gave it to his wife to card it into "slivers" by dragging it between paddles studded with wire hooks. For the remainder of the year, the women in the family spun the slivers into yarn by twisting them together on a big wheel while standing and which they turned with a wooden wheel finger held in the hand. (79-37)

The spinning wheel and the hand loom were found in virtually every home in New England.

The creation of linen began when the farmer uprooted his flax plants in mid-summer, soaked them in water until they rotted, fragmented its hard sheath on a heavy wooden flax brake, then slashed at it with a wooden blade to get out the larger chips. (79-37)

The women in the family then took the fibers and dragged them through coarse iron combs called hatchels to remove the rest of the splinters. During the rest of the year, the women spun the cleaned linen into yarn while seated at a little wheel turned by a treadle. The linen made shirts, work smocks, shifts, towels, and big napkins. Strips were sewn together to make sheets. (79-37)

Sometimes wool and linen yarns were woven together to form "linsey-woolsey," which was almost as warm as pure wool and much more durable.

Tanneries

The raising of cattle for export of salted beef to the sugar islands of the West Indies resulted in an abundance of hides. Farmers would bring their hides into town to the tanner and, through a slow process, tanners would convert these hides into leather that could be used for such products as shoes, boots, leather britches and aprons, saddles and harnesses, belts, springs for carriages, carriage tops, belts and, later, belting for textile machines.

The tanner either made leather products for sale or sold them for manufacture by others. Tanners usually kept half of the hides as payment for converting the hides into leather. They also sold the animal hair to builders for use in holding plaster together when plastering walls and ceilings. The offal from the scraped hides he sold to peddlers who, in turn, sold them to glue-makers. (79-32)

Tannery processes required large amounts of water for washing and soaking hides, and at least one tannery existed along the Quequechan River, on the south side of Bedford Street at the bottom of Rock Street (before Third Street was extended over the Quequechan River). This tanyard ("The Old Tanyard") was established in 1808 by Obadiah Chace, who operated it with his son Edmund for 75 years. The 1812 map of Fall River shows another tanyard next to the John Read house at a site on eastern side of North Main Street between Prospect and Maple Streets. There was also Reed's Tan Yard on North Main Street on a stream north of Gage's Hill (later known as Tan Pond Brook) in Steep Brook village. (64-159)

The tanner first divided the hides down the middle for easier handling and trimmed

fig. 6.10 **The tanyard**

The Obadiah Chace tanyard existed in the vicinity of the intersection of where Rock and Third Streets are now, to the rear of the main Post Office, on the Quequechan River.
Source: Tunis: *Colonial Craftsmen and the Beginnings of American Industy*

them. The hides were next soaked in water for a long period to soften them. The hides were then thrown over a slanting beam where they were scraped by a two-handled knife. On the flesh side, the knife removed fat and tissue and on the grain side the knife removed hair and the epidermis. The hides were then washed.

The next step involved soaking the hides in tannic acid. The combination of tannic acid and the gelatin in the skin toughened the hide into leather and preserved it. The tannic acid was derived from tree bark. One of the local exports from Fall River to other ports along Narragansett Bay was bark from the forests of the Copicut area, now the Southeastern Massachusetts Bioreserve. The bark was a by-product of the lumber business and was sold to tanners in the region.

Tanneries consisted of at least several soaking vats that were six feet long, four feet wide, and four feet deep. To obtain the most pliable leather, the hides were first soaked in a weak solution of bark called ooze. The tanner gradually increased the strength of the ooze over several months before the real tanning process began. Hides were then laid in the dry vats separated by one-inch thick layers of crushed tree bark. The vats were filled with water and the hides left to soak for as long as a year. (79-33)

When the hides were ready, they were transported by cart to the Quequechan River and washed. They were then placed on racks to dry. The dried leather was then thumped with a club to toughen it.

Various leathers came from different parts of the animal or from various animals. Boot and shoe soles, for example, came from the butt of a cowhide near the backbone. The thinner parts of the cowhide provided leather for boot and shoe uppers. Calfskin provided uppers for dress footwear. Skins from different animals, for example sheep and goats, required different tanning processes and had different end uses, such as gloves and bookbindings. (79-34)

One of the principal uses for locally produced leather in Fall River was for belting used for transferring power from water wheels and steam engines to textile machinery. The city's textile machinery manufacturing business—and the rapid construction of new mills—provided a market for a considerable amount of belting.

Historian Phillips says that there was a long belting shed along the shores of the Quequechan River, probably adjoining the tannery at the bottom of Rock Street.

Shoe Production

Tanners commonly made shoes and other leather products during the winter months, although the increasing demand for shoes for the growing population and for the West Indies trade resulted in the establishment of independent cordwainers or shoemakers.

One of these cordwainers, Pardon Davol, built a home and a shoe shop on North Main Street in the vicinity of what is now the North Burial Ground. He employed several persons in addition to his sons at the shop; they made shoes for the local population and for export to the Southern states and the West Indies sugar islands. Pardon's son Abner was also a shoemaker and had his own shoe shop in the West End Village in 1803. (65-13)

Phillips says that the local or itinerant cobbler would visit local farms and "he would measure the master for his boots, the mistress for her shoes, sew the tops and peg the soles, as the different branches of the trade were reached and when the work on the family shoes was completed, his yearly visit to the farm would be ended." (65-14)

Salt Production

Salt was an especially important commodity before the advent of modern refrigeration. Salt was used universally for the preservation of fish and meat. Salt cod became the mainstay of the Massachusetts economy.

The highest quality salt came from the West Indies, and one of the most important commodities in the extensive trade that emerged between New England's coastal towns and Barbados and other islands in the Caribbean was salt. This trade continued until the advent of the Revolutionary War, when the British Navy cut off trade between the American Colonies and the West India islands. The Freetown town meeting voted to send ships out to obtain salt from the islands in defiance of the blockade, but it is not known if the authorization was carried out.

As a result of this blockade, the Continental Congress directed coastal towns to begin making their own salt from seawater. This salt was inferior to the salt from the islands, but it was better than no salt at all. Seawater averages 3-4 parts of salt per 100 parts of water. Salt works soon began being established in towns all along the Atlantic coast.

In July 1777, the Town of Freetown voted to establish a salt works at the southernmost point of the town, where the Taunton River water would be most brackish. Such a salt works was established at what is now the northwesterly corner of Davol and Brownell Streets, where the Mechanic Mill now stands, just upriver from where the Quequechan River enters the Taunton. At that time, the river shoreline was probably closer to this location.

Heating the water to evaporate the liquid was considered too expensive, and a method of natural evaporation using a windmill was developed. Arthur Phillips, in his *History of Fall River,* describes that at the Davol/Brownell Streets location, a large frame was erected measuring 70-100 feet long and about 30 feet high:

> This frame was loosely stacked with fagots [large sticks] from top to bottom. The seawater was then pumped by windmills to the top of the frame and allowed to trickle slowly down over the fagots. Thus a greater surface for evaporation was obtained. The more concentrated salt solution was caught in shallow catch basins beneath the frame and then recirculated over the fagots until the salt began to crystallize out. The salt was knocked off of the fagots, and the brine was then run off into wide, shallow settling basins where the rays of the sun removed the rest of the water. The crystallized salt was worked over with rakes to secure as complete drying as possible. Sometimes a purification process was attempted but it is doubtful if this was done very often. The settling basins were provided with wood covers for protection in case of rainy weather. Of course, the warmer the climate the shorter was the time required to produce a yield of salt. (63-86)

In Dartmouth, the next town to the east, a salt works was erected in Padanaram, at the western side of the bridge that leads into that village. This salt works remained in operation almost into the advent of the twentieth century.

The 1812 map of Fall River shows a "salt house" near where the Quequechan enters the Taunton River. In 1780, the town sold the salt works for one year at public auction, the rent being ten bushels of salt, payable "in the fall."

Wood Harvesting

The 1812 map of Fall River shows a "wood slide" on the hillside north of the falls at the bottom of what is now Central Street. In his *History of Fall River,* Phillips says "there was then a wood slide through which cord wood was slid to a wharf from which it was loaded into small vessels and shipped for sale in Newport and Bristol where the local supply of wood was very limited." (63-73)

The landscape along the coast in this area was virtually denuded of forest because of the thousands of years of use of the forest for heating and cooking purposes by the native tribes. Each family kept a fire burning year-round, as did the Colonists later. Within their immediate vicinity, coastal towns found a short supply of cordwood for heating, for construction lumber, for barrel staves, and for many other uses.

While the coastal areas were mostly grassland and cultivated fields, there was extensive woodland in the interior areas above the Quequechan River falls and especially beyond "Watuppa Lake" (before the Narrows was connected with a causeway in the early 1800s, North and South Watuppa Ponds were one lake). Along the shore of the ponds, wood would be loaded on to sailboats, which would then travel down the Quequechan River and unload the wood where the Quequechan River ceased to be navigable above Main Street. From there, the wood was loaded on teams pulled by oxen and brought to the edge of the hill near what is now Central Street.

Because of the steepness of the hill and because roads were unpaved well into the 1800s, it would have been dangerous to bring the oxen teams down the hill on muddy paths strewn with ledge outcroppings. It was much safer to bring the wood to the top of the hill and slide it down to the base of the slope to a wharf where waiting boats would take it away to Newport, Warren, and Bristol.

It is likely that this slide existed for about 150 years, from about 1680 until after 1830, when schooners began bringing coal from Nova Scotia to Fall River up Mount Hope Bay and the Taunton River.

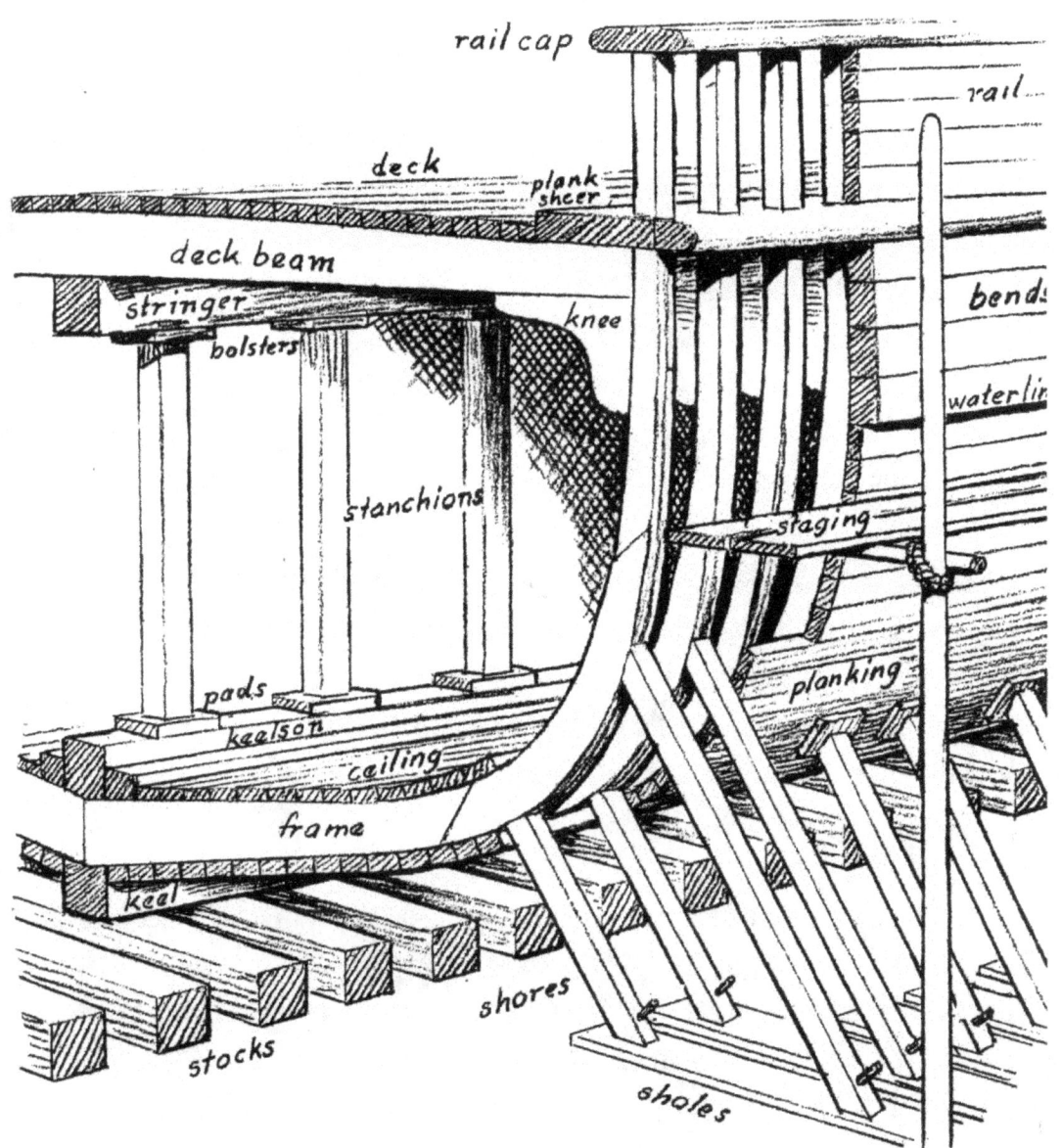

fig. 6.11 **Construction of a wooden ship**

Source: Tunis: *Colonial Craftsmen and the Beginnings of American Industry*

7 Early Iron Making on the Quequechan River

One of the more neglected aspects of Fall River's history is the legacy of iron manufacturing on the Quequechan River. In his *Centennial History of Fall River*, Henry H. Earl notes that, in 1813, there was a "blacksmithy with trip hammer" in the village of Fall River. (25-10) That reference indicates that iron was being manufactured in a forge on the Quequechan River even before the village entered the industrial age. Within a short span of eight years, this iron forging enterprise in a blacksmith shop resulted in the creation of the Fall River Iron Works in 1821. Within a few years, an observer noted in 1826 that there existed at the mouth of the Quequechan River an "extensive iron works, including one rolling and slitting mill, one nail factory, one blasting furnace and a forge." (29a-472) In addition to its own business, the Iron Works management also initiated various enterprises in cotton textile manufacturing, textile printing, railroads, and steamships.

Alfred J. Lima

The Importance of Iron in the Early American Colonies

In the early Plymouth Colony, iron implements were a rare commodity:

> The few iron cooking pots and farm tools which the Pilgrims brought with them on the Mayflower were so prized that they were bequeathed in wills along with dwellings, furniture, books and cattle. Most of their implements were made of wood. Indeed, when Governor Bradford and Captain Myles Standish proposed building a Cape Cod Canal, there was not an iron shovel in the country. Ebenezer Tomson, a grandson of Francis Cooke of the Mayflower, had a wooden shovel with a pointed iron end. It was considered so superior to their own wooden spades that it was greatly in demand by his borrowing neighbors. (17-13)

During the first years of settlement in New England, iron products were imported from Great Britain. However, the Colonists had few resources to trade for such essential items. The early discovery of bog iron in the Colonies was therefore a great boon to the settlers, and it resulted in the creation of many centers of iron making, particularly in southern New England and the middle Colonies.

By the early 1700s, deforestation in Western Europe was making iron production and its charcoal-hungry processes increasingly expensive. Because of the scarcity of wood on the British Isles, English iron production fell from 17,000 tons per annum to 12,000 tons in 1725. (40-15) As the cost of importing Swedish iron increased, Great Britain began to rely on the American Colonies as a source of pig iron. America's abundant bog iron and seemingly limitless forests for charcoal assured England an unlimited supply of pig iron for its forges.

Initially, pig iron was exported from the American Colonies to England as a raw material to be used in manufacturing iron products in Great Britain. However, as iron production and expertise improved in America, the Colonies eventually reduced their export of pig iron to England, improved their expertise in manufacturing iron products in American blacksmith shops, and generally became less and less dependent on British imports of manufactured iron products. (39-6)

Alarmed at this loss of both pig iron imports and manufactured iron exports, Parliament passed the Iron Act in 1750 "to encourage the Importation of Pig and Bar Iron from His Majesty's Colonies in America; and to prevent the Erection of any Mill or other Engine for Slitting or Rolling of Iron; or any Plateing Forge to work with a Tilt Hammer; or any Furnace for making Steel in any of the said Colonies." (35-17)

Not only did England fear industrial competition from America, but iron production also meant the ability to make swords, guns, cannon, and cannonballs. However, the restrictions of the mother country were in practice unenforceable because there was simply too much iron in America and because of the sheer number of iron-producing facilities and first-rate blacksmiths in the Colonies. Britain couldn't prevent the Colonial blacksmith from forging anything he pleased. When the American Revolution occurred, American iron smelters and forges adequately supplied the needs of the Continental Army and Navy. (87-134)

The Elements Required to Manufacture Iron

Blacksmithing in the Colonial Era required a local source of iron and that source was bog iron. Many blacksmiths located sources of bog iron in their area and built small smelters where they produced the iron necessary for their needs. Iron production had three requirements: (1) the existence of iron-rich sediments in ponds or marshes; (2) extensive forests for making charcoal fuel; and (3) clam and oyster shells that provided a lime flux. The existence of all three of these resources resulted in a thriving iron industry in the Taunton River basin. The Taunton Iron Works, the first iron manufacturing facility in the Old Colony, began operating in 1652. (17-46)

Iron ore

The iron ore used in New England for over 170 years, from 1644 to about 1820, was taken from the region's wet coastal areas, in bogs, marshes, and the bottoms of ponds and streams (hence the name "bog-iron"). The iron originates from granite rock and soils that contain iron deposits. These iron deposits are leached out by rainwater and the dissolved iron transported to streams and ponds and deposited in layers in the sediment.

> In the bog, the iron is concentrated by two processes. The bog environment is acidic, with a low concentration of dissolved oxygen. In the acidic environment of the bog, a chemical reaction forms insoluble iron compounds which precipitate out. But more importantly, anaerobic bacteria growing under the surface of the bog concentrate the iron as part of their life processes. Their presence can be detected on the surface by the iridescent oily film they leave on the water, another sure sign of bog iron. (45-2)

Because bog iron continually renews itself, it can be harvested every generation, or about every 20 to 30 years. (17-13)

Fall River's bog iron ore initially came from the Copicut area (now the Southeastern Massachusetts Bioreserve) and was probably transported to East Freetown smelters or down the Quequechan River by boat. Colonial Era iron works in East Freetown obtained their ore from local streams and swamps and from the Assawompsett Pond complex of ponds located in Freetown, Lakeville, and Middleborough.

From 1760 to 1820, Assawompset Pond alone produced 500 tons of iron ore a year. (25a-1) Bog iron was removed from the banks of slow moving streams by persons with spades on long sleds, raked from ponds, or shoveled from swamps. (76-2 and 25a-1) This iron ore was in the form called limonite. Every four tons of bog ore yielded one ton of pig iron. An average worker could harvest about one ton of ore a day. (25a-1) The wet ore had to be dried before use.

As the number and size of foundries in the Taunton River basin grew, the increasing cost of charcoal (a consequence of the depleted forests in the region) and the decreasing amount of good bog ore resulted in the importation of ore and pig iron from New Jersey, beginning about 1816. This New Jersey "mountain ore" or "rock ore" was of much higher purity than bog iron. This raw material was brought in to Fall River, Assonet and Taunton iron works by local ships on return trips. (17-49)

Later, in the 1840s, Swedish and Russian bar iron was imported to the Fall River Iron Works. (38-33) Iron from these countries made the best nails. (74-7)

Charcoal production

The second ingredient needed to make iron was charcoal, which provided the fuel and carbon required for smelters and forges. Charcoal making was the most time-consuming and expensive aspect of iron production and was produced by the distillation of wood to its carbon content. Charcoal provided the fuel for both the smelters that produced the iron and for the finery forges of the blacksmiths who shaped it. (66-1)

Charcoal was the preferred fuel for iron smelting and forging because of its high percentage of carbon (charcoal is up to 98% carbon) and because it burned hotter and cleaner than wood. Coal was known at that time, but it was rejected for blast furnaces and forges because it also was too impure. Impurities in wood and coal adversely affected the quality of the iron. For these reasons, blacksmiths used charcoal exclusively to fuel their forges. Smelters and forges in Fall River and Freetown were supplied with charcoal produced from the extensive forests of the Copicut area and East Freetown.

Charcoal was produced by burning wood slowly in a controlled, low-oxygen environment. The process of making charcoal began with the erection of a pole around which was built a wooden triangular chimney. Billets of hardwood were then stacked vertically and tightly around the chimney until a wide conical pile was formed, with a width of about 40 feet. The next step was to "slick a pit" or smooth off or even off the pile with smaller pieces of wood of from one to two inches in diameter. The pile was then covered with 1 to 3 inches of soil, clay, sod, or wood shavings combined with sand, with openings left at the top and at locations along the bottom. The chimney was then filled with kindling. (66-2)

The pile was then fired from the top of the central flue and the burn gradually spread downward and out toward the bottom of the pile. The fire had to be constantly tended so that it did not burn too quickly and turn the pile to ashes nor burn too slowly and result in an unsatisfactory burn. As the burn continued, the finished charcoal was raked from the top and perimeter of the pile.

Depending on the amount and type of wood used, a burn might last up to two weeks. One cord of wood could yield an average of 35-45 bushels of charcoal. While local farmers produced charcoal on a small scale as an extra cash crop, the operation was so delicate that larger charcoal producing works were generally left to professional charcoal burners, known as colliers. (66-2)

In transporting charcoal, it was essential that it not be excessively shaken and therefore turned into worthless dust. Therefore, it was delivered by horses in baskets or sacks or in boats on the water, as very likely occurred on the Quequechan River.

One of more important historical applications of charcoal was as a constituent of

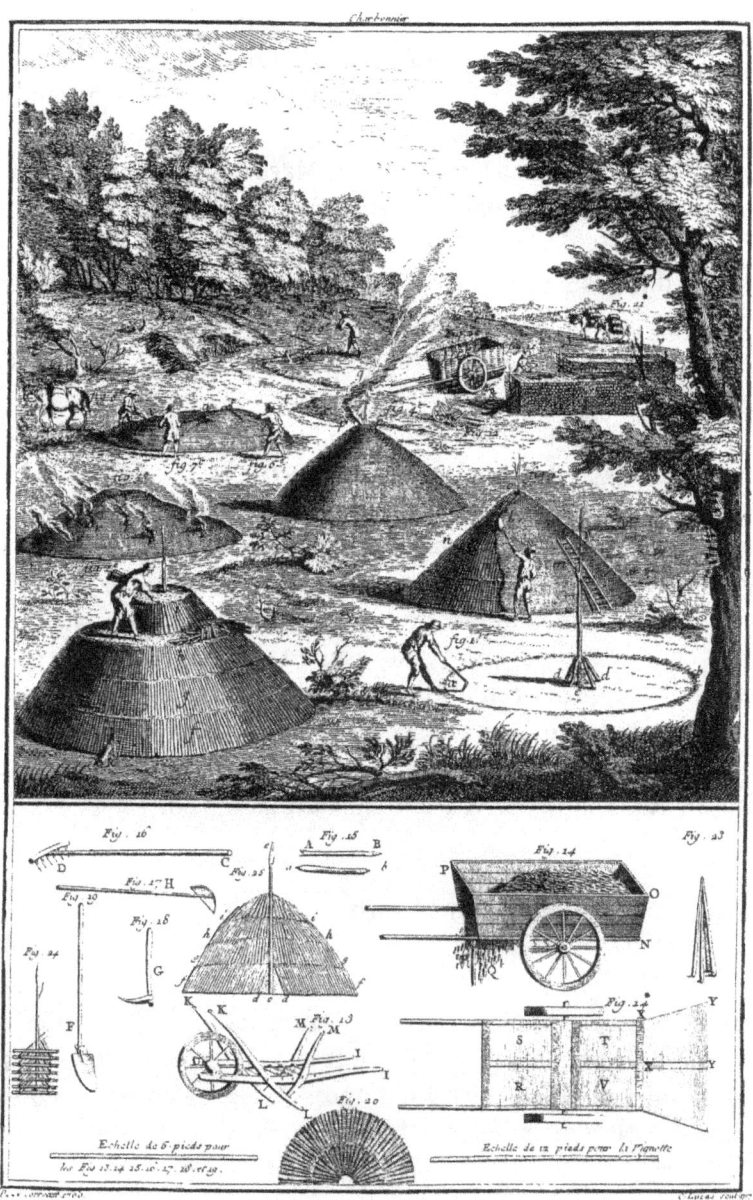

fig. 7.1 **The making of charcoal**

This 1760 French etching shows the various steps in the creation of charcoal from wood. The process began with the erection of a pole, around which stacks of wood were piled vertically in concentric circles. The finished pile was then covered with soil, clay, or sod. The collier ignited the wood from the top flue of the pile and tended the fire as it burned slowly downward under conditions of reduced oxygen. When the burn was complete (the pile would be about 50% of its original volume), the covering was removed and the charcoal carted away.

Graphic from Duhamel du Moncau, "Art du Charnonnier," *Descriptions des Arts et Metiers*, as reproduced in Hartley: *Ironworks on the Sagus*.

gunpowder, which is a mixture of saltpeter (potassium nitrate), charcoal, and sulfur. Gunpowder was produced locally for the British military and later for the militias during the Revolutionary War, one of the factors that motivated the British to raid local seaside towns.

Charcoal making continued in Southeastern Massachusetts right into the twentieth century. Personal accounts tell of charcoal making in the Copicut area of the city as late as 1920.

As local forests diminished in the northeastern coastal Colonies in the late 1700s, charcoal became more expensive. This, coupled with the scarcity of bog iron to feed the growing iron works in the area, resulted in the importation of mountain ore and pig iron from New Jersey.

About this time, coke came into general use. Coke is made from bituminous coal in the same manner that charcoal is made from wood, by heating it to high temperatures in a low-oxygen environment, a process that burns off most of its impurities. Like charcoal—but unlike coal—coke burns cleanly with no smoke and without giving off large amounts of impurities. With the introduction of the cupola furnace, which used coke or anthracite coal instead of charcoal, the use of charcoal rapidly decreased after 1850. However, coke was still not as pure as charcoal and the iron that resulted was good only for cast iron products, such as the pots and pans produced in the "hollowware" forges in Carver, Massachusetts. Coke was not pure enough to be used in the fineries, which produced higher quality wrought iron. Later English improvements in the production of iron allowed the use of coal, resulting in the exponential increase in the output of iron.

Lime flux

The third ingredient in the smelting of iron is lime flux. A fluxing agent reduces the cohesion of impurities in the iron ore, allowing the impurities to melt at lower temperatures. In the early Colonial iron works, shells of clams and oysters were added to iron smelters as a flux, which expedited the smelting process by carrying away impurities and separating slag from iron. About 450 pounds of shells were required to smelt 2,600 pounds of ore. (17-16) As the availability of shells decreased and sources of lime were discovered in the Colonies, lime was substituted as the flux.

Colonial Iron Manufacturing

From the 1600s to the mid-1800s, iron was produced in America in a two-step process. The first step involved transforming iron ore into iron in smelters. The first iron smelters were called bloomeries because the iron that was formed was called a bloom. Early bloomeries or smelting furnaces were small clay-lined shafts made of stone measuring about three feet in height and one foot in diameter. At the top of the furnace was a hole into which iron ore, charcoal and clam or oyster shells were poured, with more ingredients added as the smelting continued. As smelters grew in size, the top hole was accessed

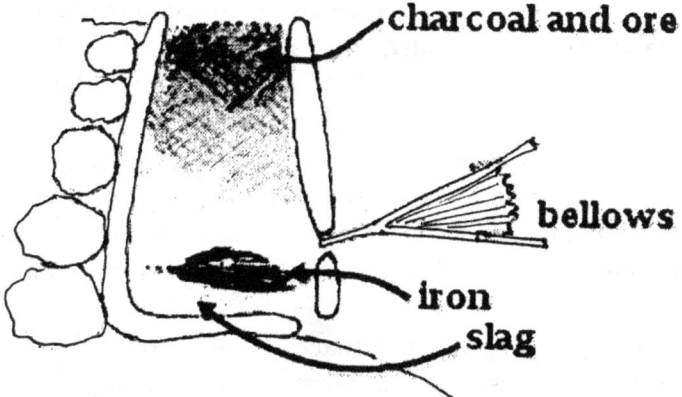

fig. 7.2 **An early smelting furnace or bloomery**

Bloomeries were small, clay-lined shafts made of stone. Iron ore was mixed with charcoal and a flux (lime or shells) and added to the top of the furnace as the smelting progressed. A bellows at the bottom of the smelter added an air blast that increased the temperature of the furnace to 2,000-2,400 degrees Fahrenheit at the bottom of the furnace. In early Colonial times, blacksmiths, needing a source of pig iron, often located sources of bog iron and built their own crude smelters to provide themselves with a reliable source of iron.

Graphic reproduced from Hurstwic: "Iron Production in the Norse Era"

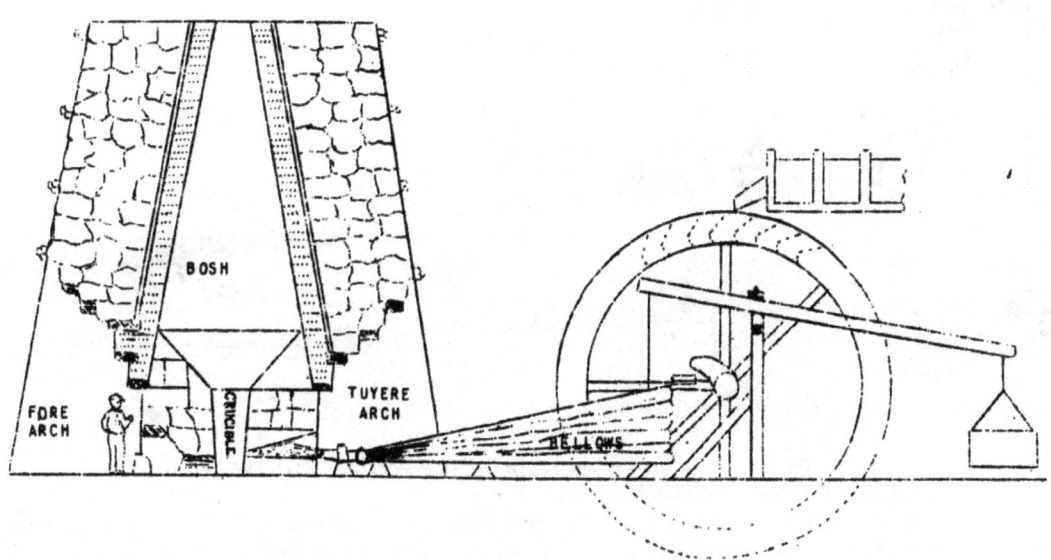

fig. 7.3 **Operation of a bellows at a blast furnace**

The arm on the wheel axle would push the bellows down, or close it, exhausting the air from the bellows. When the arm released the bellows, the weight hanging from the outer end of the tilting bar would pull the bellows up again or open it, filling it with air. Two bellows worked alternately to provide a continuous blast of air into the furnace.

Source: Murdock: *Blast Furnaces of Carver, Massachusetts*

by a ramp. Holes along the sides were provided to aid combustion, with a hand bellows providing additional oxygen for combustion. (45-3)

Bloomeries were not hot enough to melt the iron; instead, the iron collected in the bottom of the furnace as a spongy mass, or bloom, whose pores were filled with ash and slag. The lime flux improved the separation of slag from the iron. The slag waste product of the smelter fell to the bottom of the furnace and was removed via a bottom port. By repeatedly heating and hammering this bloom by hand, more slag and impurities were removed. However, this process was highly variable, inefficient and often resulted in a poor quality iron. (45-4)

Later, more sophisticated smelting furnaces—developed and spread in the Middle Ages by the Cistercian order of monks, who were skilled metallurgists—increased internal temperatures so that the ore separated into slag and molten iron. Lighter impurities and slag remained above the heavier molten iron, which fell into sand molds.

These more advanced smelters, called "blast furnaces," were able to increase internal temperatures through the use of larger bellows powered by water wheels. By forcing more air into the furnace, the larger bellows increased the internal temperature and hastened

fig. 7.4 **A waterwheel operating a trip hammer**

Henry Earl notes in his *Centennial History of Fall River* that a trip hammer was used at the blacksmith shop on the Quequechan River's upper falls. Iron forging was therefore being conducted on a small scale on the falls. Forging was done historically by a blacksmith using a hammer and anvil but, in the twelfth century, Cistercian monks began using water wheels to operate trip hammers. As furnaces increased in volume, they required more draft than human power could provide and forging the large blooms that resulted were also beyond the capability of a single man. To address this dilemma, water wheels were employed to power bellows and hammers. (85-7) The wheel turned at 35 revolutions per minute, allowing the hammerman to turn the heavy mass of red-hot iron between the downstrokes of the hammer. (5-3) The hammer pounded out remaining slag and impurities in the iron and formed it into the more regular shape of bar iron.

Graphic reproduced from: Angerstein: *R.R. Angerstein's Illustrated Travel Diary 1753-1755.*

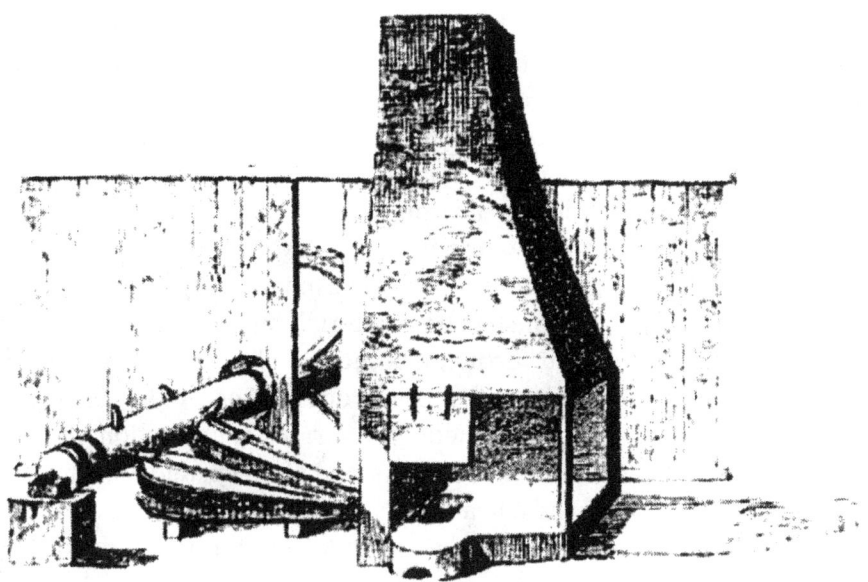

fig. 7.5 **A water wheel operating bellows at a finery forge**

A finery forge was a water-powered mill where pig iron was refined to produce wrought or bar iron. Water wheels operated bellows, which blasted air into the hearth, keeping the fires sufficiently hot. Forging of iron in a finery forge resulted in metal that was stronger than cast or machined parts because, during forging, the metal's grain flow changed, making it stronger and more ductile. The final product of the finery forge was bar iron. There were three hearths in an iron forge: two finery hearths used to melt the iron and one chafery hearth used for forging the iron into bars. In the finery, the finer re-melted pig iron to produce a lump of iron (with some slag) known as a bloom. The finery fire had to be kept at a temperature that would allow the slag to run off in a molten state but would keep the iron in a pasty condition so that it could be molded and lifted off for the hammerman. This was consolidated by a water-powered trip hammer and returned to a second finery hearth. The hammerman then took the bloom and drew it out into bar iron on the trip hammer. In doing so, he heated the iron in a third hearth, the chafery forge. The fuel in the finery and chafery forges had to be charcoal, since impurities in other fuels would affect the quality of the iron.

Graphic reproduced from: Angerstein: *R.R. Angerstein's Illustrated Travel Diary 1753-1755.*

the smelting process. The iron that melted out of the pile and settled to the bottom of the smelter was called "pig iron." Pig iron received this name from the shape of the iron that emerged from the sand molds, which looked like a line of suckling piglets on a sow. (84)

During the Colonial Era, smelters could be found in many sizes, from small furnaces dug into pits to structures up to 20 to 30 feet tall. These furnaces had to be tended continuously day and night. Workmen typically slept next to the furnace on cots.

Since the pig iron was full of slag and other impurities and was brittle, it had commercial value only for cast iron products. To produce a higher-quality and malleable iron, pig iron therefore had to be processed in a finery forge, the second step in producing iron. Forging involved the reheating of pig iron and hammering it until the metal's grain flow changed and the iron became stronger and more ductile or workable. In the finery, the finer reheated the pig iron and brought it to a huge water-powered hammer (a trip hammer) that pounded out the slag in the iron. The finery fire was kept hot enough to allow slag to run off in a molten state but at a temperature that would keep the iron in a pasty state, allowing it to be moved and lifted off to the hammer. (5-6)

The iron was then given to the hammerman, who reheated the iron in a chafery forge and then pounded out the remaining slag in the iron. He then formed the iron into a bar, known as bar iron. This forging produced the workable iron known as wrought iron. Wrought iron was then used by blacksmiths to form into various farm implements, tools, ship hardware, horseshoes, household items, and construction materials such as nails, latches, and hinges.

If the iron was to be processed further, it went to a slitting mill, whose rollers were powered by water wheels. The slitting mill formed rods or sheets to be sold to blacksmiths, who then worked them into iron components for various construction and shipbuilding purposes. Rods were used by nailers, blacksmiths, and often farmers who formed the rods into nails by giving them a point and a head.

Each ton of bar iron produced required a ton and a quarter of pig iron and three loads of charcoal. (5-2)

The Beginnings of Iron Manufacturing on the Quequechan River

Iron manufacturing on the Quequechan River evolved out of a shipbuilding enterprise initiated by two remarkable men, individuals who were to have a major impact on the industrial growth of Fall River as an iron making and textile center.

When Major Bradford Durfee joined Colonel Richard Borden's grist mill business sometime after 1812, the Major brought with him a knowledge of shipbuilding learned during his sojourn as a ship's carpenter in New Bedford. This led to a partnership between the two men to construct vessels at the mouth of the Quequechan River for the coastal and West India trades. The two men built one ship a year in their shipyard. Early in the day, Richard and his brother John would go into the Copicut woods to bring out timber, knees, braces, and other wood for the shipbuilding operation. Major Durfee and his assistants would then work up the wood into finished components for their vessels. (25-44)

However, the vessels also needed many iron components. At the end of the day, Major Durfee and Colonel Borden "would spend a good part of the night in a blacksmith shop near by, executing the necessary iron work" for their ships. (25-44) This involved heating and hammering pig iron into wrought iron in a forge and working the wrought iron into finished iron ship parts. "Working along in this way for a few years, the field and facilities for a larger business soon developed themselves, especially in the working up of iron into spikes, bars, rods, and other articles of constructive use." (25-44)

The making of iron parts for sailing vessels in the local blacksmith shop therefore evolved into a more ambitious business of processing pig iron into bar iron, which, in a few years, resulted in the formation of the Fall River Iron Works in 1821. In all of these operations, the Quequechan River powered the water wheels that operated the bellows for both the blast furnace and the forge, raised the forge's trip hammer, and turned the rollers of the slitting mill. The Iron Works' first buildings were built on the lower falls of the Quequechan River, where the Metacomet Mill is now located, near the corner of Davol and Anawan Streets. The growth of the Iron Works is discussed in more detail in Chapter 11.

8 The American Revolutionary Battle on the Quequechan River

During the Revolutionary War, the British systematically burned coastal towns that sided with the Colonies as a means of demoralizing and weakening the resistance of the rebels. Local merchant ships were especially decimated. The British held Newport and therefore secured the inland coastal areas on Narraganset Bay, Mount Hope Bay, and the Taunton River. However, certain anti-British towns such as Fall River and Tiverton were making gunpowder (from local charcoal), manufacturing small arms, and forming militia companies of Minutemen, and these rebellious towns therefore represented potential support for Continental troops against a possible land operation against Newport.

Colonel Joseph Durfee, who later started the first cotton mill in Fall River, was appointed Captain of a company of Minutemen. They guarded the shores of Mount Hope Bay and the Taunton River and prevented British ships captained by Wallace, Asque, and Howe from landing their troops. Fall River's Minutemen under Colonel Durfee also marched to cover the retreat of American troops from New York and later joined with other troops and took an active part in the Battle of White Plains. After the British took possession of Newport, the Fall River Minutemen covered the retreat of Colonel John Cook and his troops from Aquidneck Island at the Bristol ferry (now the site of the Mount Hope Bridge). This operation was hazardous because of the small boats used as ferries at the time and because of the many British spies on the island.

At Fall River, Colonel Durfee commandeered a store at the end of the wharf, at a site that later became the Fall River Iron Works, for use as a guardhouse. Sentinels were stationed there every evening, with orders to shoot if an approaching boat did not respond after being hailed three times. Colonel Durfee later related his recollections of the battle of Fall River, as quoted in Henry H. Earl's *Centennial History of Fall River, Massachusetts*:

> Not long afterward, Samuel Reed, acting as guard, discovered boats cautiously approaching the shore and receiving no answer fired upon them; that the whole neighborhood was soon in arms; that from behind a stone wall the guard kept up a rousing fire upon the enemy until they brought their cannon to bear upon us and fired grape shot among us, whereupon a retreat was ordered.
>
> Two of the guard were sent to remove the planks which has been laid over the lower stream for people to cross upon, and that we retreated slowly until we reached the main road where the bridge now crosses the stream [North Main Street where it crosses the Quequechan River] and there we formed and gave battle, whereupon the enemy retreated leaving one dead and another bleeding; that they carried away their wounded; that before the soldier who was left behind expired he said that there were 150 British in the charge commanded by Major Ayers; that upon landing the enemy set fire to the new house of Thomas Borden, then to a grist mill and saw mill belonging to Mr. Borden, both of which stood at the mouth of the fall river [Quequechan River]; that he saw them set these buildings on fire.
>
> That in their retreat they set fire to the house and other buildings belonging to Richard Borden, who was then an aged man and took him prisoner, but they were pursued so closely that the buildings which they had set fire were saved; that we continued to fire upon them as they passed down the bay; that they ordered Mr. Borden to stand up in the boat so that we would desist firing upon them but he refused to do this, and threw himself upon the bottom of the boat; that while lying there a shot killed a British soldier standing by his side; that Mr. Borden refused to answer questions and in a few days was dismissed on parole; that this engagement was on Sunday, May 25th, 1778; that they buried the two [British] soldiers near the south end of the Massasoit factory. (25-198)

During the later part of 1778, the militia procured supplies and built flat-bottomed boats and scows to prepare for crossing to Aquidneck Island with the intention of dislodging the British from Newport. The militia was joined by troops led by Lafayette, Greene, and Sullivan and marched down the island toward Newport, where they were to be joined by the French fleet. In 1778, Lafayette stayed at the home of Judge Thomas Durfee. The house has been moved several times and now stands at 94 Cherry Street. That dwelling, built in the Federal style in 1750, is named for its famous visitor and for Judge Durfee, who helped finance the Revolution.

Colonel Joseph Durfee is buried in the North Burial Ground on Brightman Street, as are Thomas Durfee and Robert Irving, two other Revolutionary War soldiers. Other soldiers of that war that are buried in the city are Captain James Simmonds and Benjamin Weaver, buried in the North Steep Brook Burying Ground, and Ephraim Boomer, Elisha Caswell, and Benjamin Peck, buried at Oak Grove Cemetery. (31-16)

fig. 8.1 **The Lafayette-Durfee house, 94 Cherry Street**

9 Fall River's Growth Into a Major Textile Manufacturing Center

Fall River's Development Within the Context of the Industrial Revolution

The development of Fall River remarkably parallels the progression of the Industrial Revolution. Before 1813, the economic life of the village of Fall River closely resembled the economic life of antiquity, thousands of years ago. The core of life was the family, where subsistence agriculture was the basic unit of the economy. Families essentially provided everything for themselves: shelter, food, and clothing. Certain families specialized in crafts and trades that served farm families, such as milling flour, tanning, shoe making, or coopering. Since the beginning of recorded time, cloth was made by women in the home by spinning wool, flax, or cotton on a spinning wheel and then weaving those threads on a hand loom.

Where cities existed, they were small in size and developed around shipping and trading with other ports. They were therefore almost always located on protected harbors on the coast or on navigable rivers.

As in antiquity, travel on land was by animal power or on water by sail.

Since the introduction of agriculture into the earliest civilizations 10,000 years ago, everyday life had certainly improved but had not materially changed for the average person. The Industrial Revolution changed all that.

The Industrial Revolution had its beginnings in the European Renaissance and in The Age of Reason and The Enlightenment that followed it. These movements emphasized the use of empiricism, reason, and science and encouraged systematic thinking, scientific inquiry, and experimentation. The Scientific Revolution that began with Copernicus and Newton inspired a transformation in physics that led to new ways of thinking about matter and, eventually, to advances in technology. In England, the rise of a merchant class, a strong trading system, population increases, and various social transformations made that country especially receptive to the acceptance of the mechanization of work.

The Industrial Revolution was not a revolution in the sense that violent change occurred overnight. However, in terms of historical time, it happened very rapidly. The hallmark of the Industrial Revolution was the application of power-driven machinery to manufacturing, which included the introduction of technological innovations that (1) speeded up work and (2) did the work of several laborers simultaneously. To do this, work had to be taken out of the home and placed in the factory. The factory provided a centralized place where large amounts of power were available to operate machines. As factories multiplied, they created a new phenomenon, the industrial city.

The Industrial Revolutions occurred in two stages. The first stage evolved slowly but with accelerating rapidity from about 1750 to 1850 and was powered by the water wheel. The second phase occurred very rapidly after 1850, following the widespread use of steam power.

The Industrial Revolution began with the mechanization of the textile industry and with improvements in the making of iron. While historians often disagree on what caused the Industrial Revolution, all agree that it started in England.

The Industrial Revolution and the development of textile manufacturing

From antiquity to the present, the manufacture of textiles has involved two basic processes: spinning, where raw fibers are drawn out and spun into yarn, and weaving, where the yarn is interlaced at right angles to one another to form a fabric.

Textile production was the first industrial activity to be mechanized, and it became the model for organizing human labor in factories. As long as cloth was made in the home by the use of the spinning wheel and hand loom, the amount of textiles produced would be limited. Because flax and cotton required time-consuming pre-processing (when they were available), the preferred cloth in Europe was wool, whose raw product could be obtained from family-owned sheep. Most European cotton and cotton cloth came from India.

It took ten persons to prepare yarn for one woolen weaver and three persons to prepare yarn for one cotton weaver. While spinners were busy, weavers were idle while they waited for their yarn.

The first innovation in producing textiles was created by an Englishman, John Kay, a Lancashire loom fixer, who patented his flying shuttle in 1733. This improved the productivity of hand weaving three-fold. However, now weavers were even idler, because they needed three times the amount of spinners than before to keep them busy. Something had to be done to increase the amount of thread produced by spinners to keep up with the hand weavers.

In 1767, the solution was at hand, when James Hargreaves, a Lancashire weaver, built his spinning jenny, a major milestone in the history of the mechanization of textile production. Hargreaves' spinning jenny was a hand-cranked machine that allowed one operator to spin 16 threads simultaneously. However, Hargreaves' spinning jenny had a major disadvantage, since the machine was suitable for producing only soft filling or weft yarns. Its twist was not strong enough for warp use. Warp thread runs along the length of a bolt of cloth and weft thread runs at right angles to the length of cloth.

If one event must be chosen as the beginning of the Industrial Revolution, it would be the opening of the textile mill of Richard Arkwright in Cromford, Derbyshire, in 1771. Capitalizing on a series of earlier inventions, Arkwright built the first machine that could draft, twist, and wind cotton simultaneously in separate zones of the same machine. The thread produced by Arkwright's spinning frame was therefore strong enough to be used for the long warp threads. This was the first time that warp cotton threads could be produced by machines, a major breakthrough in the production of machine-produced all-cotton cloth.

Arkwright's other and equally major contribution to revolutionizing the production of yarn was to adapt his spinning frame so that it could be powered by mechanical means. After experimenting with horses, Arkwright decided that water wheels were the best method to power the machines, thus earning his invention the name of "water frame." Under Arkwright's direction, Cromford became the first cotton textile factory town.

The invention of Hargreaves' spinning jenny and Arkwright's spinning frame caused a demand for speeding up the pre-processing of cotton for spinning, such as carding and roving. In 1775, Arkwright improved on an earlier invention and took out a patent for a new mechanized "carding engine."

In 1778, Samuel Crompton of Bolton, Lancashire, invented a semi-automatic version of the spinning jenny that could be powered with a water wheel. Crompton's "spinning mule"—which produced large quantities of weft or soft filling yarn—finally ended the shortage of yarn. Arkwright's spinning frame and Crompton's spinning mule became the twin pillars of spinning, with Arkwright's frame producing the strong warp thread and Crompton's mule producing the soft weft yarn. (24-11)

Indeed, the proliferation of new cotton textile factories using Arkwright's spinning frame and Crompton's spinning mule now produced too much yarn. Now, the problem became one of finding enough weavers to weave the yarn that was being produced. Since there were only so many home hand weavers in the English countryside, a way had to be found to mechanize weaving, a much more complex process than spinning.

Edmund Cartwright, an English minister, finally met this challenge in 1785. Cartwright's power loom was driven by drive shafts and was another key invention of the Industrial

fig. 9.1 **The spinning jenny**

Developed in 1767 by James Hargreaves, the spinning jenny increased from one to 16 the amount of spindles a single person could operate (this would later increase to 80). However, the machine had the disadvantage of producing only soft filling or weft yarn and was powered by hand. Later, in 1778, Samuel Crompton would invent his "spinning mule," which improved on the spinning jenny and applied it to water power.

Graphic source: Merrimack Valley Textile Museum

Revolution. While Cartwright's power loom was a breakthrough, it performed poorly, and it took several other inventors another 25 years before the process of power weaving was perfected.

The full development of cotton textile manufacturing, however, was stymied by yet another bottleneck: cotton tended to be expensive and available in limited quantities because the separation of seeds and leaves from the raw cotton required many hours of intensive hand labor. However, that problem was resolved by the invention of the cotton gin in 1793 by an American, Eli Whitney. The cotton gin mechanically separated the seeds from the cotton and, with the aid of a water wheel, could clean much greater volumes of cotton and with much more speed than by hand. This not only made cotton cheaper but resulted in its widespread cultivation. The way was now open for the massive expansion of the cotton textile manufacturing industry.

The development of new chemicals, and in large batches, was an essential breakthrough in the Industrial Revolution. In the bleaching of textiles, for example, the bleaching of cloth had been done since antiquity by soaking the cloth in alkali or sour milk followed by repeated exposure to the sun in bleach fields. With the development of bleaching powder (calcium hypochlorite) by Scottish chemist Charles Tennant about 1800, the bleaching of textiles was reduced from months to days.

Since the early cotton manufacturing machines were wooden with cast iron parts, the development of more efficient and faster spinning and weaving machines that were all metal had to await improvements and innovations in iron and steel manufacturing and milling. That finally occurred in the early 1800s, allowing the development of such advances as the Horrocks power loom.

For 50 years, the industry expanded until a new bottleneck emerged, and that was the finite number of fall sites on rivers and the maximum amount of power that could be obtained through water wheels. This was overcome by the invention of a steam engine that could rotate a wheel, developed by Scotsman James Watt in 1763. This invention, and its improvement in 1849 by George Corliss of Providence, Rhode Island, assured that the growth of the textile manufacturing industry was limitless.

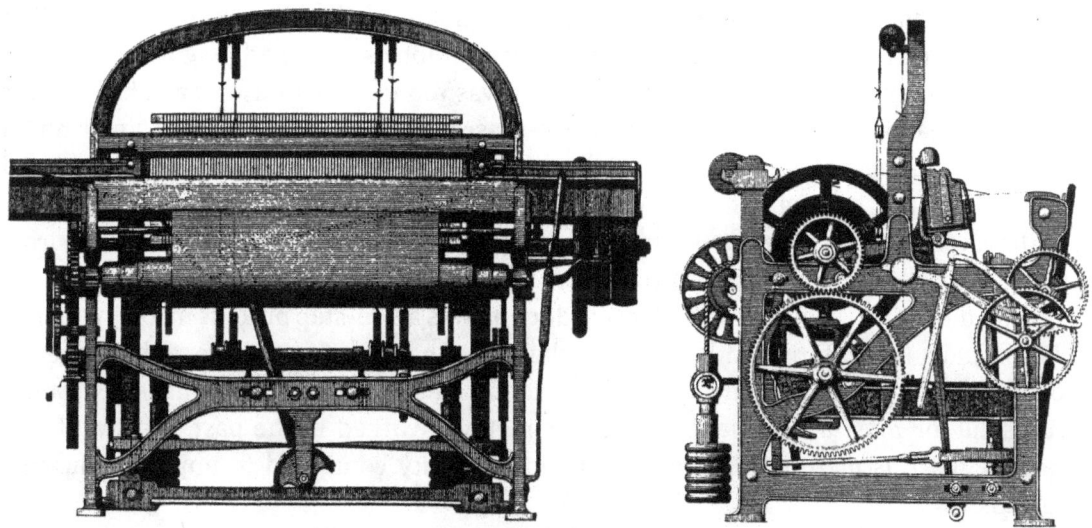

fig. 9.2 **Horrocks power loom**

When Edmund Cartright developed a workable power loom in 1785, it performed poorly. It took 25 years before the process of power weaving was perfected in the loom developed in 1813 by William Horrocks and later modified by him in the model above, about 1835. The development of the Horrocks loom, however, had to await the invention of steel processes and measuring instruments that would allow the creation of milling machines for the mass production of other machines. Before the development of milling machines, looms were created one at a time by blacksmiths or mechanics.

Graphic source: Baines: *Cotton Manufacture*

One remaining restriction to textile growth was that clothes were made by hand at home or in shops using a thread and needle, the method of making clothes used since ancient times. This slow method limited the use of cloth. The development of the sewing machine in 1847 by a Massachusetts mechanic, Elias Howe, mechanized sewing. However, the mass production of clothing (and the expanded use of cloth) became possible only with a more practical sewing machine developed in 1851 by Isaac Singer of Boston, whose invention was the first really practical, domestic sewing machine that had a straight needle, a foot treadle, and the ability to sew curved seams.

Iron, steam, and the Industrial Revolution

The Industrial Revolution could not have occurred without improvements in the making of iron and steel. For thousands of years, iron had been made in small bloomeries, using bog iron ore, charcoal, and limestone. The only major advancement in the making of iron was in the Middle Ages, when Cistercian monks began using water wheels to power bellows. This power source allowed the construction of larger "blast furnaces."

The Industrial Revolution in England resulted in a growing demand for iron (and charcoal), resulting in the depletion of Britain's forests. The best steel could be produced only from pure ores available from Sweden and Russia; however, pig iron imports purchased from cartels in these countries were becoming increasingly expensive. The situation demanded a workable solution that did not rely on imports. Coal had long been available as a fuel for making iron in Great Britain, but it was too impure to use as a substitute for charcoal. A partial resolution to this problem was developed in 1709, when Englishman Abraham Darby successfully used coke in making iron. Coke is a purer form of coal created in a manner similar to making charcoal from wood.

However, Darby's process was not the solution, since his end product was cast iron, which was too brittle for most uses. Coke was still not pure enough.

Until the Industrial Revolution, iron production was a two-step process involving a blast furnace and a finery forge, both fueled by charcoal. The blast furnace created impure pig iron, and the finery forge refined the pig iron to remove impurities and to make it stronger and useful for most applications. While this method worked in the past, the scarcity of charcoal and this method's limited production capacity continued to impede industrial progress.

That changed in 1784, with the introduction of the "puddling" furnace developed by British ironmaster Henry Cort. The puddling furnace finally freed iron making from the constraints of charcoal. Now it was possible to convert cast iron into finer wrought iron in large batches using coal—and without the use of the ancient finery forge, which now became obsolete. The way was now open to use two of Great Britain's abundant natural resources—coal and iron ore—to produce high quality iron in large amounts for industrial use.

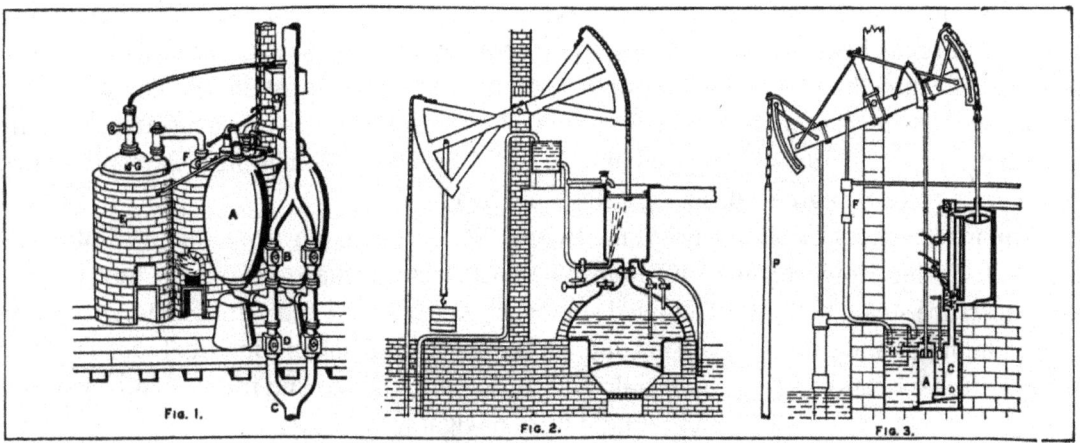

FIG. 1. SAVERY'S STEAM ENGINE (1698). FIG. 2. NEWCOMEN AND CAWLEY STEAM ENGINE (1705). FIG. 3. THE WATT PUMPING ENGINE.

fig. 9.3 **The early steam engines**

Source: *Encyclopedia Britannica*, 1940 edition

The evolution of steam power

The growth of the iron industry was hampered by limitations on coal production from mine flooding. This occurred when seepage water gradually flooded the coalmines as the mines deepened, limiting their life span and yield. This restriction on coal production limited iron production and, therefore, the growth of the young industrial economy. Various devices were in use to pump water out of mines, most using animal power. However, in 1698, Englishman Thomas Savery, using a design developed earlier by Edward Somerset, introduced a rudimentary pump powered by steam that he called the "Miner's Friend." While it was a breakthrough in being the first use of steam for industrial purposes, it was slow and its lift was limited to less than 25 feet.

Thomas Newcomen, an English blacksmith, introduced the first machine in 1702 that begins to look like a modern steam engine. Newcomen used Savery's unmodified pump but added a piston to provide direct mechanical power to run external pumps. Newcomen's "atmospheric engine" allowed water to be lifted to any height and is credited with allowing mines to be made twice as deep as before. However, Newcomen's engine was still very slow and required substantial amounts of coal to operate.

While repairing one of Newcomen's engines in 1763, a Scotsman named James Watt began to experiment with improvements to the machine. Watt's modifications made Newcomen's atmospheric engine a reciprocating engine, thereby transforming it into a true "steam engine." Even more importantly, Watt added a crank and flywheel to provide rotary motion. This innovation transformed steam machines from simply pumps into devices that could power machinery in mills. While Watt's engine quadrupled the power of Newcomen's engine, it still produced only 5-10 horsepower.

The steam engine, more than any other machine of its time, required advanced machinery and metallurgy to be truly efficient. The further refinement of the steam engine into a more powerful machine had to wait until the development of hardened steel that could be used in building machines that could make other machines. One of these breakthroughs was the invention of the cylinder-boring machine in 1775 by ironmaster John Wilkinson, which made Watt's steam engine a practical source of power.

The massiveness of the early steam engines restricted them to stationary industrial uses in factories. However, in 1799, a Cornish blacksmith, Richard Trevithick, developed a high-pressure steam boiler that allowed steam engines to be portable enough to be used on rail locomotives and steamboats. This opened the way for the development of railroads and the easy transportation of raw material and finished products to and from inland locations. Industrial cities could now be sited inland from navigable waterways.

Subsequent improvements to the Watt engine, particularly those introduced by George Corliss in 1849, increased its power and reliability until the steam engine was widely adopted after 1850, ushering in the second phase of the Industrial Revolution.

The Introduction of Textile Manufacturing into America

Official English Colonial policy suppressed manufacturing in the American colonies. The role of the colonies was to provide the mother country with raw materials and to import manufactured goods from England. Following the American Revolution, an ongoing controversy began on whether the new country should be a nation of farmers or whether manufacturing should be introduced and encouraged.

Thomas Jefferson was the champion of an agrarian nation of yeomen farmers, for he felt that a land-owning nation of tillers of the soil was the bedrock of democracy. Jefferson thought that America should "let our workshops remain in Europe." Alexander Hamilton, on the other hand, advocated that manufacturing was the only way to achieve a strong country and to free the United States economically from England. In the 1780s, 50 percent of all English exports went to America. Hamilton won out, and industrial associations began providing grants and bounties to support local experiments and encourage industrial espionage in England. (24-11)

Since England kept close guard on its new technologies, only a general knowledge was available of the Arkwright water frame. Between 1786 and 1790, repeated attempts to replicate the Arkwright spinning machine met with only limited success. However, the simpler spinning jenny was known in this country and a dozen "manufactories" sprang up spinning yarn using men or animals to turn the wheels of the jenny. Meanwhile, as attempts to import the Arkwright design failed, Yankee ingenuity focused on making the non-working models operative.

Rhode Island was the center of experimentation with the Arkwright model. At this point, the enthusiasm for textile manufacturing infected Moses Brown, a wealthy Quaker merchant who came out of retirement to take up the challenge of spinning cotton into thread using waterpower. He established a business making cotton cloth for his son-in-law, William Almy, and cousin, Smith Brown, and set up business as Almy and Brown in a Providence marketing house. He converted his assets into cash and quickly went about buying most of the available textile equipment then available in the area. The factory included hand-powered spinning jennies and hand-powered looms in one location. Because the thread produced by the spinning jennies was only suitable for the weft, stronger linen was used for the warp. However, the factory was not much better than those that preceded it. (24-13)

Continued experimentation with Arkwright-type machines continued to be unsuccessful.

Brown hired the talented Quaker blacksmith, Ozil Wilkinson, to build prototypes of the Arkwright spinning frames, but none worked. Undaunted, Moses Brown advertised that he was eager to employ anyone from England who had any knowledge about Arkwright's water frame. One of those who answered was Samuel Slater of Derbyshire, who had worked in a mill with Arkwright machines and had memorized its key features. Within two months of arriving in Providence under the employ of Moses Brown, Slater constructed a hand-

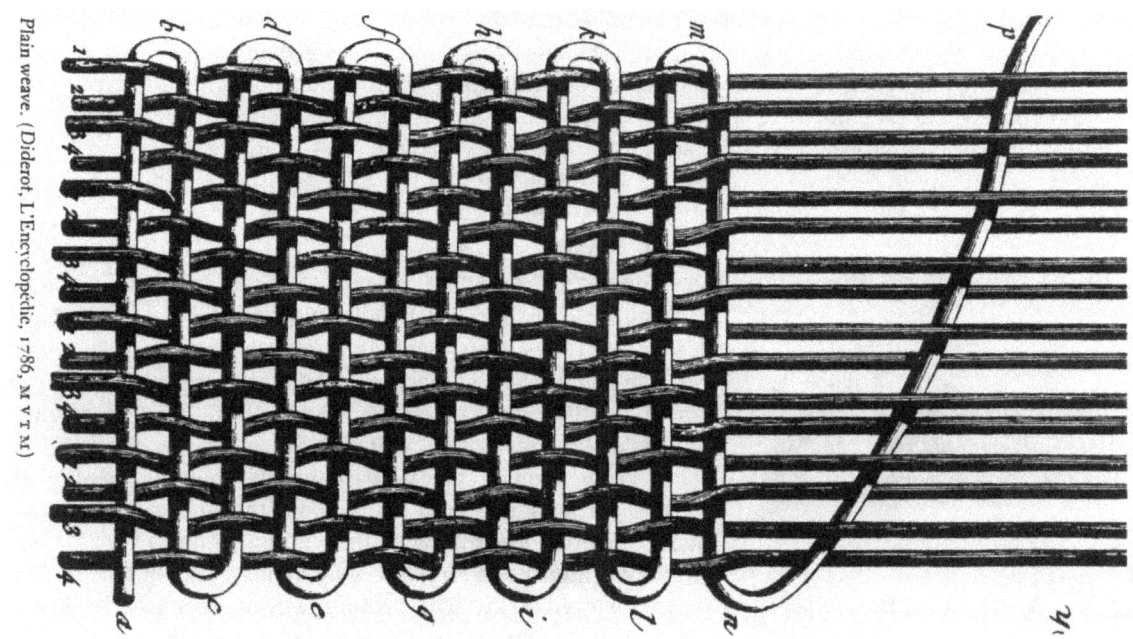

fig. 9.4 **Warp and weft**

The horizontal strands of thread shown in this etching are the "warp" yarn, or the stronger threads that continue lengthwise in a bolt of fabric. The vertical strands of thread shown are the "weft" yarns that are woven at right angles to the length of a bolt of cloth. The weft threads are softer and less strong than the warp threads and are therefore filler yarn.

Graphic source: Diderot: *L'Encyclopedie,* 1786

Moses Brown

Raised in Providence, Rhode Island, Moses Brown (1738-1836) was a wealthy merchant and farmer and shared the family business with his brothers Nicholas, Joseph, and John Brown. The brothers came from Baptist roots and were founders of Brown University. When Moses Brown became disillusioned with the family's slave trade, he broke away and became a Quaker. He subsequently retired and became an ardent abolitionist and advocate for the poor. In 1784, he founded the Moses Brown School, one of the nation's oldest preparatory schools. He shared the prevailing Quaker belief that manufacturing might relieve social ills.

fig. 9.5 **Moses Brown**

In 1789, he came out of retirement to establish his son-in-law, William Almy, in a profitable business and chose textiles. After Brown advertised widely that he would set up in business anyone with a knowledge of the Arkwright process, an Englishman named Samuel Slater presented himself as someone who knew about the secret manufacturing process. After Slater made good on his promise, Moses Brown financed a partnership of Slater and Brown relatives in a mill on the Blackstone River in Pawtucket.

The Brown family and their relatives were investors in the early Fall River textile mills and the Fall River Iron Works through their connections with the Borden family, particularly Holder Borden and Jefferson Borden. Holder Borden became the agent for the Blackstone Company, owned by Brown and Ives and, in 1830, convinced Nicholas Brown and Moses Brown Ives of Providence to invest in the Massasoit Mill and to manufacture cotton goods there under the name of Brown, Ives, and Borden. Jefferson Borden's connections were through marriage and friendship while managing the Iron Works interests in Providence.

Samuel Slater

Rightly called the "Founder of the American Industrial Revolution," Samuel Slater began learning spinning as a 14-year-old apprentice in the mill of Jedediah Strutt, the partner of Richard Arkwright. He was intelligent and observant and, in three years, became an overseer of machinery and mill construction. At the age of 21, disguised as a farmer, he passed the English customs and sailed for America in 1789.

Hearing from a packet captain that Moses Brown was looking for someone who knew about the Arkwright process, Slater negotiated terms with Brown and came to Providence. After two months in Brown's employment, Slater had built a hand-powered Arkwright prototype from memory. Within a year following, he had harnessed the system to a water wheel.

fig. 9.6 **Samuel Slater**

Moses Brown financed the partnership of Almy, Brown, and Slater and set the firm up in the Slater Mill. At his death in 1835, Slater was a rich man and owner of 13 textile mills.

There is a direct line between the Slater Mill and the new mills at Fall River, since David Anthony worked for Slater for four years before he built the first textile mill on the Quequechan River. What Anthony learned working under Slater he applied to the operation of his Fall River Manufactory. After a few years working for David Anthony, Holder Borden moved to Pawtucket to work overseeing a mill for the Wilkinson brothers, in-laws of Samuel Slater.

powered version of the Arkwright machine. In March 1790, Almy and Brown reported that they were manufacturing strong warp thread that now allowed them to create an all-cotton cloth.

However, other challenges needed to be overcome, including the demanding task of developing a perpetual cylinder card and harnessing the machine to water power. Nonetheless, in December 1790, just over a year after Slater arrived in Providence, the system was complete and the lumbering water wheel on the Blackstone River in Pawtucket began to turn the gears that set the frames and cards in motion. Slater's success earned him a partnership with Almy and Brown, in a new manufacturing venture, Almy, Brown, and Slater. The Browns and Slater began to build on their success with new mills on the Blackstone River. (24-14)

> By the turn of the nineteenth century the primary features of the Rhode Island factory system were already established: a small riverside mill equipped with cylinder cards and Arkwright spinning frames, children working within, minimal and crowded housing nearby, and a company store. Raw cotton was cleaned by hand. Yarn was either "put out" to be woven on consignment by hand weavers or was hand woven in the mill itself. In both cases, the finished cloth was sold by the merchant-manufacturer. (24-15)

The Rhode Island and Waltham textile Manufacturing Systems

New England developed two distinct models of textile manufacturing: the Rhode Island system and the Waltham system. Fall River and the Blackstone River Valley mills are the classic examples of the Rhode Island system while Lowell, Lawrence, and Manchester are the ultimate representatives of the Waltham system.

In *The Run of the Mill*, Steve Dunwell describes the Rhode Island and Waltham manufacturing systems as differing in the following ways:

Rhode Island System	Waltham System
Mills developed on streams of less than 1,000 horsepower, sometimes as low as 100 horsepower.	Mills developed on water power privileges of more than 1,000 horsepower, sometimes much higher.
Concentrated in Southern New England, particularly along the Blackstone River Valley in Rhode Island and Massachusetts and elsewhere in New England where smaller mill privileges could be found.	Concentrated at falls along the major rivers of Northern New England.
Ownership of mills by resident families and local partnerships.	Absentee ownership by joint stock corporations.
Relatively small capitalization.	Total investment was very large.
Development was typically one mill or a cluster of mills, independently owned, that evolved organically over time.	Development was in pre-planned cities with large complexes in mills.
Management was by the owners of the mills.	Management by agents of the owners.
Distribution of water power arranged independently by each mill.	Water power distributed through a system of canals, and shared by several mill corporations, leasing their power from a canal company.
Children and families preferred.	Young women preferred.
Family housing and tenements preferred.	Boardinghouses preferred.
Small operations emphasized diversity of product line, selling direct to markets.	Mass production was emphasized, with sales through marketing agents.

With the development of steam power and increasing immigration, the distinctions between the Rhode Island and Waltham systems began to lessen over time. (24-52)

Early Cotton Textile Manufacturing on the Quequechan River

In 1803, the year that Fall River was incorporated as a separate town from Freetown, there were only 18 dwellings in the village of Fall River. By 1812, the village had grown to 30 dwellings—three saw mills, four gristmills, one fulling mill, two blacksmith shops, and several small stores.

fig. 9.7 **The Colonial Joseph Durfee mill at Globe Corners**

Source: Fenner: *History of Fall River, Massachusetts*, 1911

Prior to 1812, farm families produced their own wool or linen, which was then spun in the home into yarn on foot-powered spinning wheels. The yarn was then woven into cloth on hand looms. The wool typically came from sheep grown on each farm. Farm families were self-sufficient economic units and provided for their own basic needs.

The devastating effect on New England's important maritime commerce from Thomas Jefferson's Embargo Act of 1805—coupled with the reduction of imports from England as a consequence of the War of 1812—resulted in a search for new commercial activities by New England investors. The growing success of textile manufacturing in England became the impetus to create a textile industry in America, particularly along the major rivers in Northern New England and the smaller streams in Southern New England, including the Quequechan.

The first cotton-spinning mill to be built in Fall River was erected in 1811 by Colonel Joseph Durfee on a stream and pond at what is now the intersection of South Main and Globe Streets, at Father Kelly Park. At the time, this area was in Tiverton, Rhode Island.

Colonel Durfee's mill was a cooperative venture between factory and cottage. The process began with cotton being distributed to farm families to conduct the labor-intensive task of picking it—that is, removing seeds, leaves, and stems from the raw cotton. The cotton was then sent to Colonel Durfee's mill (and to the early mills on the Quequechan), where it was spun into yarn. The yarn was then sent back to the farmers' cottages to be woven into cloth on hand looms. The cloth was then returned to the mill for finishing and readied for sale.

The finished goods were subsequently carted two miles to Fall River village where they were sent by schooner to Providence and other local markets. The cloth was also sold over the counter at the mills' company store.

Before the introduction of reliable water-powered looms, Pawtucket's Samuel Slater depended on more than 600 cottage hand weavers within a 60-mile radius of his mills to weave the output of his spinning mills into cloth. "The mills in the neighborhood of Providence kept wagons running constantly into the rural districts, invading both Massachusetts and Connecticut, bearing out yarn to be woven and returning with the product of the hand-looms, worked by the farmers' wives and daughters of the country side." (25-78) Hand weavers could not keep pace with yarn produced by the spinning mills. This "putting-out" system paid as low as two cents a yard to the home weavers. (25-29)

In his 1877 *Centennial History of Fall River, Massachusetts*, Henry H. Earl mentions that this method of cottage "outsourcing"—where cotton was brought to the farmhouses for picking, brought back to the mill for spinning into yarn, brought back to the farmhouses for weaving, then brought back to the mill again for finishing—took a considerable amount of

David Anthony

Born in Somerset, Massachusetts, on January 9, 1786, David Anthony can be said to be the father of cotton textile manufacturing in Fall River. At the age of 14, he began employment with John Bowers, a wealthy Somerset merchant, where he assumed increasing responsibilities in various aspects of the Bowers business. In 1808, at the age of 24, he joined Samuel Slater at his power cotton spinning mill in Pawtucket. With the knowledge gained in Pawtucket, Mr. Anthony joined with Dexter Wheeler to form the first textile manufacturing mill on the Quequechan River in the village of Fall River. In 1817, he also installed the first power looms on the Quequechan in the Fall River Manufactory. David Anthony retired from active business in 1839 but continued as the director of other textile mills and various banks. In his seventies, he came out of retirement to assist in organizing Union Mill No. 1. He died at 81 on July 6, 1867. His counsel to young men was "Happiness and success in a business life are promoted by correct habits, systematic living in all matters, and great promptness in fulfilling engagements." (25-13) Graphic source: Earl: *Centennial History of Fall River, Massachusetts*.

fig. 9.8 **David Anthony**

bookkeeping to keep track of who had what at any one time. Earl quotes an earlier source as saying:

> In the cotton business of that day there was a great amount of book-keeping and clerical work, of which very few manufacturers now have any idea. Every bale of cotton put out to be picked was booked, as was also every web given out to be woven. A mill of seven thousand to ten thousand spindles required more labor to take care of the yarn after its leaving the reel and prepare if for or get into the market, than all the spindles in Fall River now [1859] demand. (25-18)

The machinery in the Durfee mill, built completely of wood, was constantly breaking; Colonel Durfee never made a success of the operation, which closed in 1829. Colonel Durfee died a poor man in 1843.

The success of the textile industry in Fall River began with the mills that began appearing on the Quequechan River in 1813. In that year, two textile mills were erected on the Quequechan River. The first to begin operations was the Fall River Manufactory, which opened in October 1813. David Anthony and Dexter Wheeler organized it, both experienced in textile manufacturing at other locations. David Anthony, born and raised in Somerset, had worked for four years (1808-1812) with Samuel Slater, the father of American textile manufacturing, at Slater's mill in Pawtucket. From that work, Anthony learned about the workings of the Arkwright machinery. Anthony then associated with Dexter Wheeler for a year at Wheeler's horse-powered cotton spinning mill in Rehoboth. Wheeler was experienced not only as a manufacturer but also as a blacksmith and mechanic.

The power available on the Quequechan River drew Anthony and Wheeler to Fall River. The Fall River Manufactory mill measured 60 feet by 40 feet, was three stories high and took only seven months from the conceiving of the building to the start of operations. Its first story was of stone and the upper two stories of wood. It was the first cotton spinning organization on the Quequechan River and contained 1,500 spindles. It was built on the third falls above tidewater. (25-12)

The second mill built on the Quequechan River was the Troy Manufacturing Company, whose name was shortly changed to the Troy Cotton and Woolen Manufactory. This mill was larger than the Fall River Manufactory, measuring 108 feet long by 37 feet wide and four stories in height. It was designed to run 2,000 spindles. The Troy mill was located at the top of the falls and was made of stones gathered from neighboring fields. It began operations in March of 1814. (25-15)

Two smaller buildings accompanied the construction of the Troy Mill: a two-story machine shop and a blacksmith shop with two forges. Both were rented to John Borden Jr., originally from Aquidneck Island. Mr. Borden came to Fall River in 1813 via Waltham, where he and his two brothers worked in the machine shop there. He probably acquired his knowledge of cotton spinning machinery from his Waltham experience, since that community was the site of the first cotton mill operated by Francis Cabot Lowell. (25-15)

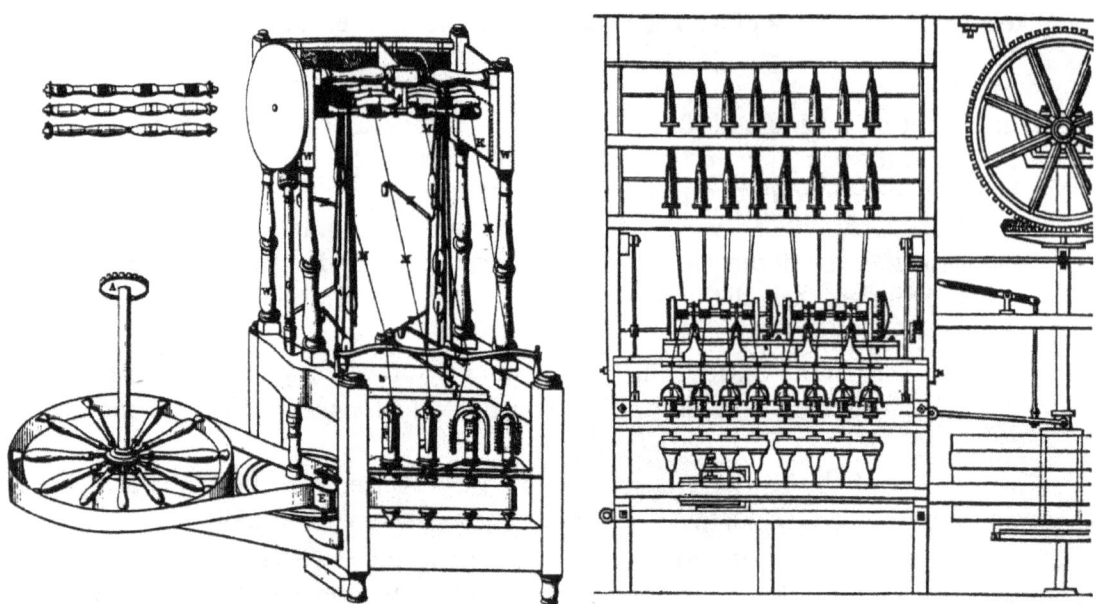

fig. 9.9 **The Arkwright spinning frame**

Richard Arkwright first began using his water-powered spinning frames in England in 1771. However, it wasn't until Samuel Slater, in an act of industrial espionage, memorized the design and built the first prototype in America in 1790 did the Industrial Revolution begin in the United States. In another feat of industrial espionage, David Anthony, formerly employed in the Slater Pawtucket mill, brought the design of the Arkwright spinning frame to Fall River, where it powered the first successful cotton textile mill on the Quequechan River.

Graphic source: Baines: *Cotton Manufacture*

Having a blacksmith and machine shop connected to the mill was a common characteristic of early textile mills, since the cotton spinning machinery (and later looms) were built and repaired on-site.

In 1813 and 1814, the Fall River Manufactory and the Troy Cotton and Woolen Manufactory built the first tenements in Fall River. These several buildings were four-family structures that also housed the mill agent. (25-19)

The originator and agent of the Troy mill was Oliver Chace, a carpenter and wheelwright, who began experimenting with power cotton spinning at a small mill in Dighton. Chace was the only one of the 28 original stockholders who had any practical knowledge of how to operate a mill. (25-16)

A third mill, the Union Cotton Factory, was erected in 1813 on the same stream (at Globe Corners) as Dr. Durfee's mill in what was then Tiverton, Rhode Island. Edward Estes and others organized that mill.

At first, these mills operated in the same manner as Colonel Durfee's mill, where cotton was put-out to farm families during various phases of the process. However, Francis Cabot

Lowell eventually secretly imported a design for a water-powered loom into America. He perfected the design, created a loom made entirely of wood, and installed it in his first mill in Waltham in 1814. (24-31)

In 1817, four years following their start-up, the first power weaving began in the Fall River Manufactory on three power looms manufactured by Dexter Wheeler. These early looms were very heavy and clumsy and were constantly getting out of order, weaving one yard of good cloth and ruining the next through inadequate control of the shuttle. The dressing was very poor, and at times the yarn would mildew and rot on the beam, causing large quantities to be thrown away. (25-16)

When power looms improved, they were introduced into the Troy mill in 1820.

The third company to be incorporated on the Quequechan River was the Pocasset Manufacturing Company, which built a mill on the upper falls of the river. The Pocasset Manufacturing Company was organized in 1822 with $100,000 in capital. Its purpose was not to manufacture textiles itself but to provide space for rental for new small businesses. The new Hawes, Harris Machinery Company occupied two stories of the building and Miller Chase operated a grist mill in the basement. The Pocasset Company continued to add to the property as the demand increased for such leased space. The principal owner was Samuel Rodman of New Bedford, whose family was prominent in New Bedford's whaling industry.

The market for the early Fall River cotton cloth was quite distant from the village. The major portion of the yarn and cloth from the Fall River Manufactory was sold in Philadelphia and the products of the Troy mill were sold in Massachusetts, New Hampshire, and especially Maine. (25-18)

The end of the War of 1812 and the reduction of tariffs in 1816 resulted in an economic depression and the resumed importation into America of textiles from England. These events caused a sudden decline in the infant New England textile industry. Just as the new cotton mills began operation in Fall River, cotton cloth dropped in price by 50 percent. As a result, others did not follow the first two mills on the Quequechan for several years.

In 1816, Francis Cabot Lowell and Nathan Appleton, founders of the Waltham textile system, visited Pawtucket, Rhode Island, and found that "all was silent, not a wheel in motion ... all was dead and still." (24-32) However, Lowell lobbied Congress to pass protectionist tariffs in 1824 and 1828, which resulted in a return to prosperity for the early textile mills and for the country as a whole.

From 1820 to 1830, with the imposition of tariffs to protect against the importation of cheap English goods, industrial activity on the Quequechan River grew at a rapid pace. The early mills constructed additions to their original buildings and new mills were built on the falls. These included the Pocasset Manufacturing Company, the Annawan Manufacturing Company, the Massasoit Mill, the Robeson Print Works, the Satinet Factory, the Nankeen Mill, and the complex of the Fall River Iron Works at the lowest falls. (25-22)

A writer in 1827 noted that Fall River was an incongruous "city of the wilderness, rising in the midst of hills, trees and waterfalls and rural scenery." That same writer marveled at the pace of frenzied activity in the town and remarked that "industry is the presiding goddess of Fall River; an idle man could no longer live there than a beetle in a bee hive." (29a-475)

In 1833, as the village began to evolve into an industrial town, Samuel Slater visited the area and described the site in this way:

> Situated on a rather abrupt elevation of land rising from the northeast side of Mount Hope Bay, distant about 18 miles from Newport and nine from Bristol, R.I., stands the beautiful and flourishing village of Fall River, so called from the above river, which, taking its rise about four miles east, runs through the place, and, after many a fantastic turn, is hurried to the bay through over beds of rock, where, before the scene was changed by the hand of cultivation and improvement, it formed several beautiful cascades, and has a fine imposing effect. The village is now only picturesque from the variety of delightful landscapes by which it is surrounded: the background presenting a variety in rural scenery—where neat farms and fertile fields show themselves here and there between hill and dale, and rock and wood. (65-8)

fig. 9.10 **Calico printing, about 1836**

Textile roller printing, an inexpensive alternative to block printing, was introduced into America from England in 1823. Roller printing allowed much higher volumes of cloth to be printed than was possible with block printing. Successful roller printing required precision machining and highly skilled operators and engravers. American Print Works recruited experienced craftsmen from Scotland and England.

Graphic source: Baines: *Cotton Manufacture*

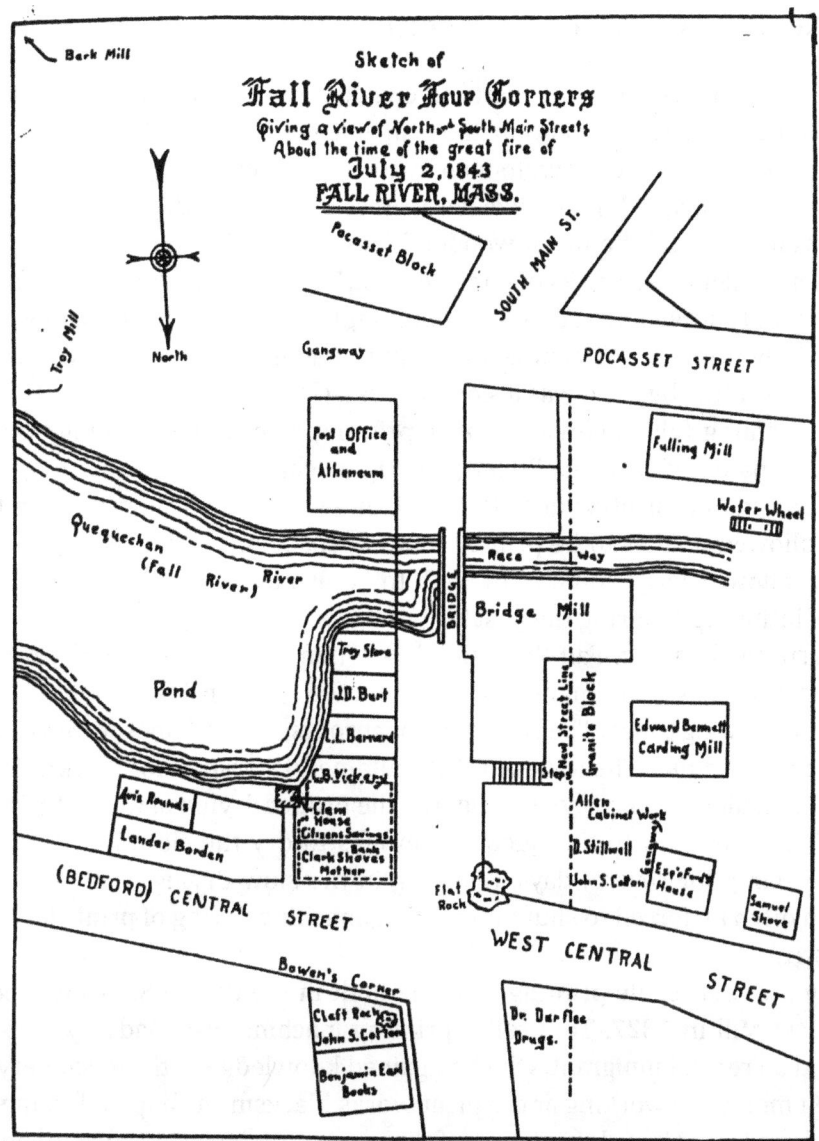

fig. 9.11 **Fall River Four Corners, 1843**

This sketch shows the area around the Quequechan River above the falls immediately before the great fire of 1843, which destroyed the center and a substantial part of the town. The Bridge Mill is Samuel Rodman's first Pocasset Mill, built in 1821. The water wheel brought water up from the river below to the surface for housewives to use for household use. The fulling mill, which processed woven cloth and made it usable, was at that location since Colonial times and remained in use well into the industrial era until it was destroyed in the fire. Edward Bennett's carding mill is below the Bridge Mill. The Post Office and Athenaeum was where the Skeleton in Armor was exhibited and where it was also destroyed in the fire. Dr. Durfee Drugs was the store of Dr. Nathan Durfee, the Harvard educated doctor who built a mill in the Copicut area of the city to make dyes for the textile mills. That mill's walls are still standing in what is now the Southeastern Massachusetts Bioreserve.

Map source: Phillips: *The Phillips History of Fall River, Fascicle I*

Early Textile Printing on the Quequechan River

For thousands of years, from the ancient civilizations of China, India, and Egypt to the early years of the Industrial Revolution, the principal method of printing textiles was block printing. This was conducted by craftsmen, who carved out a design in relief in blocks of wood, with the design standing out like letterpress type. A length of rolled cloth was then spread across a table and the inked wooden blocks pressed against the cloth and struck with a wooden mallet to ensure a good impression. The cloth was then rolled forward over drying rollers and succeeding sections of cloth impressed with the blocks until the length of cloth was fully printed. Different colors required additional runs through the blocking process. Block printing by hand was a slow process. (85a-3)

Machine printing (also known as roller printing or cylinder printing) was invented in England and patented in 1785. Roller printing mechanized the printing operation and allowed the continuous printing of cloth from engraved copper rollers. The first machines of this type allowed textiles to be printed in up to six colors at a single operation. Roller printing now allowed large volumes of cloth to be printed, and it has been the principal method of printing cloth during the past 200 years. (85a-6)

Textile printing began in Fall River in 1824, when Andrew Robeson of New Bedford, related by marriage to the Rodmans of that city, rented space in Samuel Rodman's Pocasset Mill and began printing calico. The early printing process in the Robeson Mill consisted of a hand and block technique. The Robeson Mill contained 100 tables for block printing. (25-31)

One of the challenges of early printing was the proper drying of the cloth. Large drying sheds were built adjacent to the Quequechan falls to dry the cloth in the air. When the weather was damp for several days, resulting in the slow drying of the cloth, the print works would often be forced to shut down. Later, machine drying of print cloth solved this problem. (25-31)

Probably the first textile printing machine built in the United States was constructed in the Pocasset Mill in 1827. That roller printing machine was made by Ezra Marble of Somerset and a French immigrant, who had gained knowledge of the machinery in France. The two men met while working at the print works blacksmith shop and, combining their efforts, developed and altered the design of the machine, which continued in operation for many years. (25-30)

Business grew rapidly, and in 1826, Robeson built his own factory on the Quequechan falls. The buildings for Robeson's Fall River Print Works increased in number until the last and largest was completed in 1836. Etched copper rollers were introduced in 1832 and yard-wide rollers in 1837. (25-30)

In 1827, only 42 of the 700 persons employed in the village were recorded as "foreigners": English, Scottish, and Irish immigrants. Most of these new arrivals worked in the print shops as block printers and later as copper roll engravers. (10-48) Block printing continued until 1841, when a strike resulted in all of the block printers being fired and

fig. 9.12 **Commerce on the Taunton and Quequechan Rivers, 1877**

replaced with machine roller printers. (25-31) The centuries-old tradition of block printing had come to an end on the Quequechan.

At the Globe Corners, the cotton mill established by Colonel Joseph Durfee was converted to a print works in 1829. Holder Borden operated a print works there in 1833-34 and it passed through various ownerships, including W. & G. Chapin and the Bay State Print Works, after which it was purchased by the American Printing Company. (29a-566) Near the print works, off of Globe Street, there still stands is a double row of Cape Cod cottages on Chapin Street. These cottages were built for immigrant Scottish block printers who were employed at the print works.

Fall River's Advantages as a Textile Center

The Quequechan River was an ideal location for the development of the textile industry during the era of water power for several reasons. First, it had an adequate flow of water for powering mill machinery, but the flow was also modest enough to protect against flooding. Second, its bed was granite, allowing a firm footing for mills to be built directly over the river with their water wheels set in the stream bed. Third, the water privileges of the Quequechan were owned by a few descendants of Joseph and Richard Borden, who, by 1714, had gained ownership of the entire stream. (73-20)

The Quequechan River, with its eight falls, was large enough to operate mills of the Rhode Island type, yet small enough to be developed by modest amounts of capital available along the Taunton River. The river was too small to be attractive to Boston capitalists, who were interested in the large horsepower privileges available along the major rivers of New England, such as the Merrimack River, the Connecticut River, and the rivers of northern New England. (73-38)

In addition to the reliable flow of water of the Quequechan River, its sound bedrock foundation, and its ownership in one family, the location of the river next to a navigable waterway, the Taunton River, made Fall River an ideal site for textile manufacturing. It was the combination of water power and coastal location—not one or the other—that accounted for the ascendancy of Fall River as a textile center. (73-38)

Significantly, Fall River's location between Boston and New York and its proximity to Long Island Sound made it the best natural water privilege at tidewater anywhere in New England south of New Hampshire.

Local capital was used to build the first several textile mills in the city, coming principally from Fall River and from families in surrounding towns that made their fortunes during the Colonial Era and later in the shipping business. Approximately 58 to 82 percent of the capital for these first textile ventures came from the towns of Somerset, Freetown, Dighton, and Fall River. (73-29)

Later, larger water-powered mills on the Quequechan River required more capital, and that came from New Bedford (from the profits of the whaling industry) and Providence. New Bedford had no water power and would not become a textile center until the advent

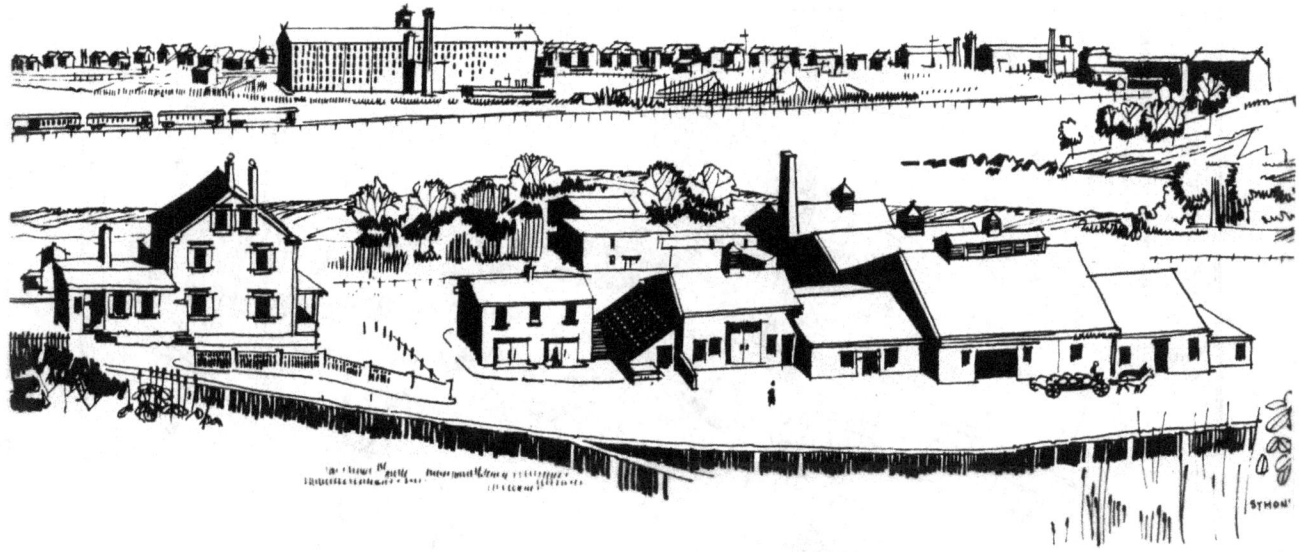

fig. 9.13 **The Quequechan River valley about 1877**

Looking south from Pleasant and Quarry Streets.
Source: David Symons in *Fall River 2000*

of steam power. In 1821, Holder Borden and his relatives convinced Providence capitalists to invest in the Iron Works and, later, textile mills. (73-34)

In these early years, New Bedford investors became associated with the development of textile mills at the upper falls of the Quequechan, and Providence investors became associated with the development of the lower falls at the Iron Works and their spin-off commercial ventures. Later, however, industrial growth in the city was principally funded through local investors from the profits of earlier textile and iron manufacturing ventures.

Thomas Russell Smith, in his book *The Cotton Textile Industry of Fall River*, states that Fall River textile mills had three advantages over mills north of Boston:

1. Transportation costs for coal and cotton were somewhat lower than interior points or to the New Hampshire and Maine coasts;

2. Climatic conditions favored the production of finer goods in coastal Southern New England because of the region's higher relative humidity and more even temperature; and

3. Fall River's location convenient to major cloth markets in New York City, Philadelphia and Boston and to print works in Rhode Island, New York, and Pennsylvania. (73-38)

Alfred J. Lima

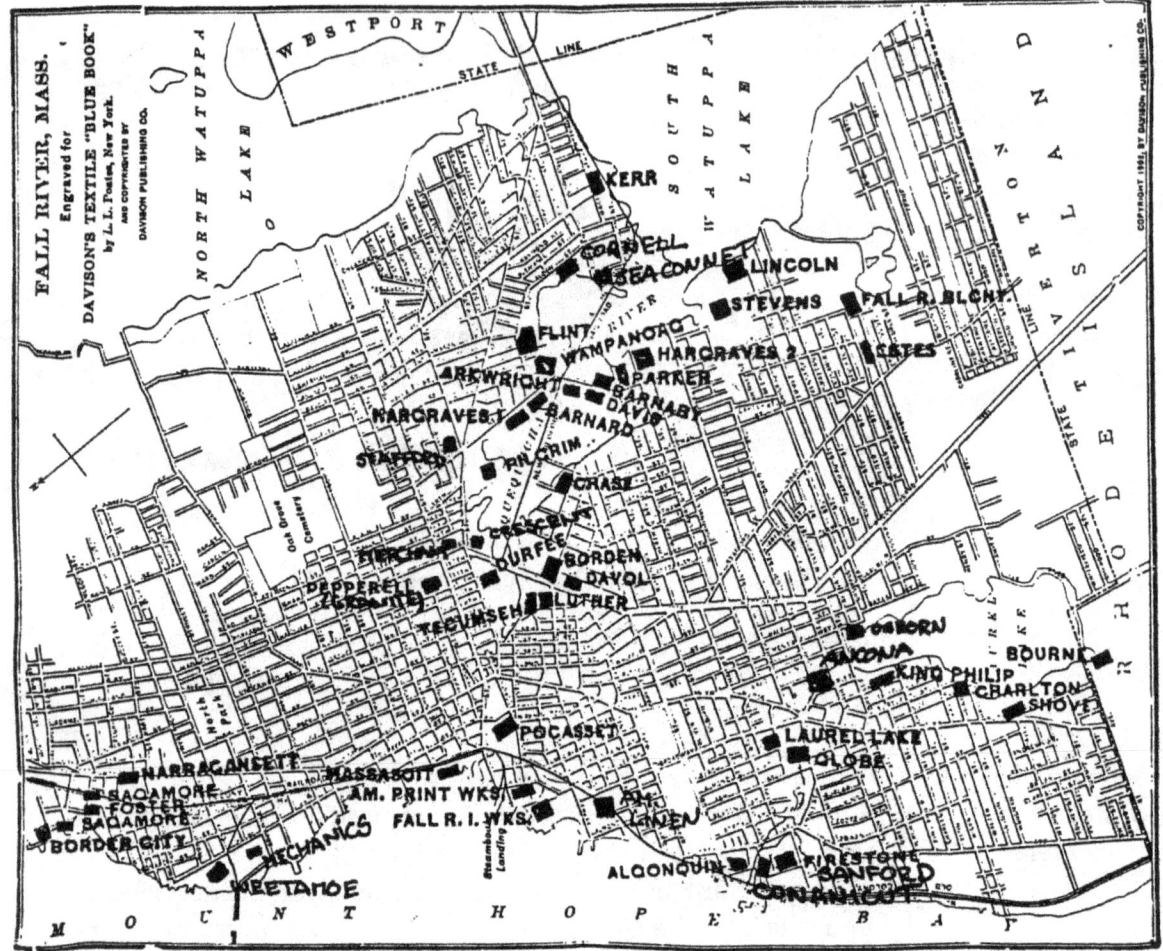

fig. 9.14 **Location of textile mills in Fall River, 1912**

Source: Archives of the Fall River Public Library

The fact that Fall River was located on a navigable waterway, the Taunton River and Mount Hope Bay, with immediate coastal access, meant that coal could be delivered from the mines directly to the city entirely by water, reducing costs. Coal schooners brought coal from Maryland and West Virginia via Potomac River ports, then later from Norfolk and Newport News.

South of Cape Cod, coal delivery costs were considerably lower than north of the Cape because the dangers and time involved in rounding the Cape were avoided. Rail charges for the coal deliveries to Lowell and Lawrence were higher than for Fall River. Because of cheaper coal transportation to the city, the costs of manufacturing textiles in Fall River was 13 percent lower than in the inland Merrimack River centers of Lowell, Lawrence, and Manchester, NH. (73-54)

Another advantage for Fall River was that the Fall River Iron Works owned coal mines in Frostburg, Maryland. This and other mining investments became very profitable for their Fall River owners. The direct access to cheaper coal contributed to the city's competitive advantage.

Fall River's locational advantage also applied to raw cotton, which was brought by ship directly from Savannah, Georgia and Charleston, South Carolina. The city's location south of Cape Cod resulted in cheaper shipping costs for cotton.

Climate also played an important role in textile production in the city. Fall River's climatic conditions are similar to the English textile region of Lancashire. The city's location south of Cape Cod and on the Taunton River and Mount Hope Bay provided a higher relative humidity and more even and milder temperatures than other textile centers. Only in the late 1880s did artificial humidifiers become generally adopted.

Fall River's location south of Cape Cod with immediate access to Long Island Sound allowed it to be nearer to finished cloth markets and to finishing centers, all accessible by water transportation. New York City was the premier selling market for textiles. With the rise of the Fall River Line, and its freighting capacity, orders could be placed at Fall River mills on one afternoon and arrive the next morning by boat in lower Manhattan, convenient to the city's textile merchandising and garment districts. Few textile centers could match this service. (73-63)

Fall River's textile production was in finer printed cloths, and the city's mills were able to accommodate the printing of calico cloth before 1855. After that time, however, the rapidly increasing growth of the mills in the city outstripped local printing capacity. Print works in Rhode Island, New York City, and Philadelphia were easily accessible to Fall River manufacturers via water transportation.

The Basic Processes of Cotton Textile Manufacturing

In his book *Natural and Man-Made Textile Fibers*, George E. Linton describes how cotton undergoes a transformation from a farm crop to fine thread. The creation of cotton textiles begins with the planting, harvesting and ginning of cotton. Only one-third of the harvested cotton crop is cotton fiber, the remaining two-thirds being waste matter such as leaves, pieces of pods, chaff, and other material. First-time ginning takes place at the plantation cotton gin or at the "community gin." It is then brought in bulk form to the major cotton cities of the South, where it undergoes a second ginning process and is compressed into cotton bales of about 500 pounds. (53-34)

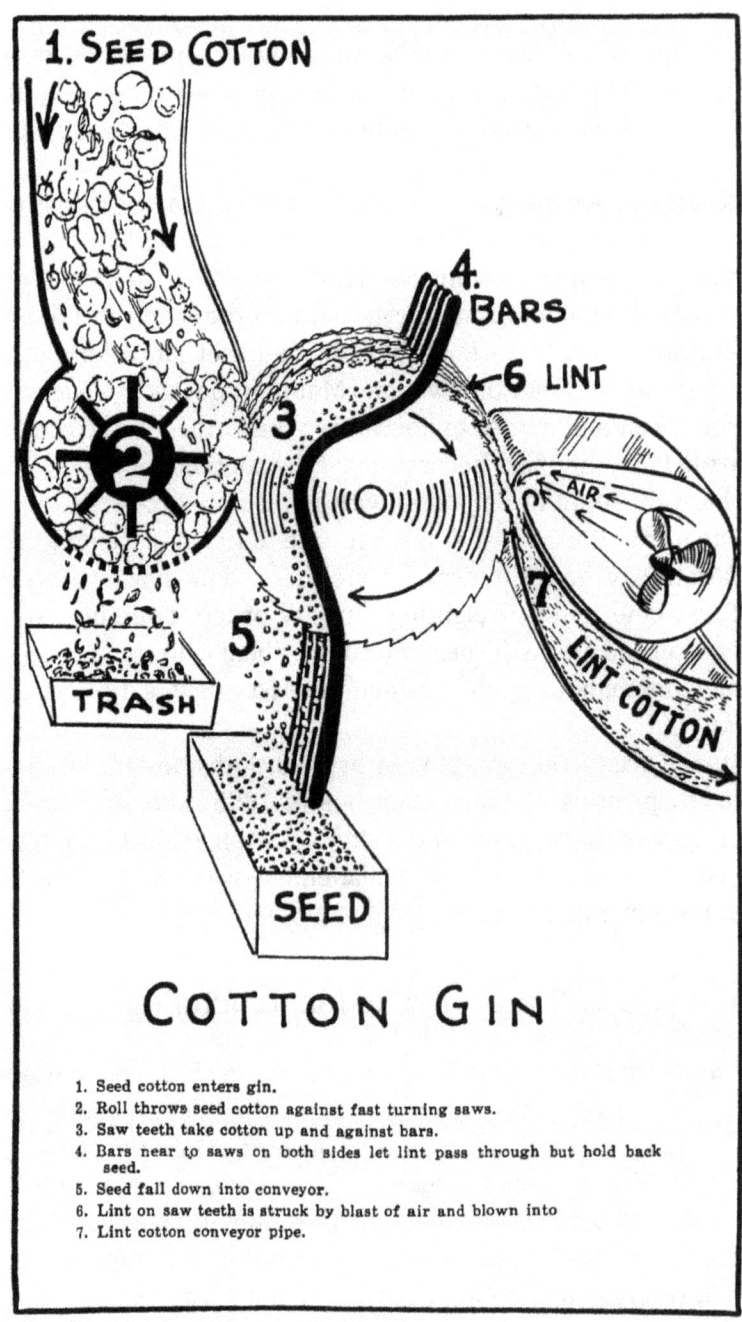

fig 9.15a How raw cotton is transformed into thread in a textile mill

From the Bibb Manufacturing Co., Inc. Macon, Georgia, as shown in Linton: *Natural and Manmade Textile Fibers*

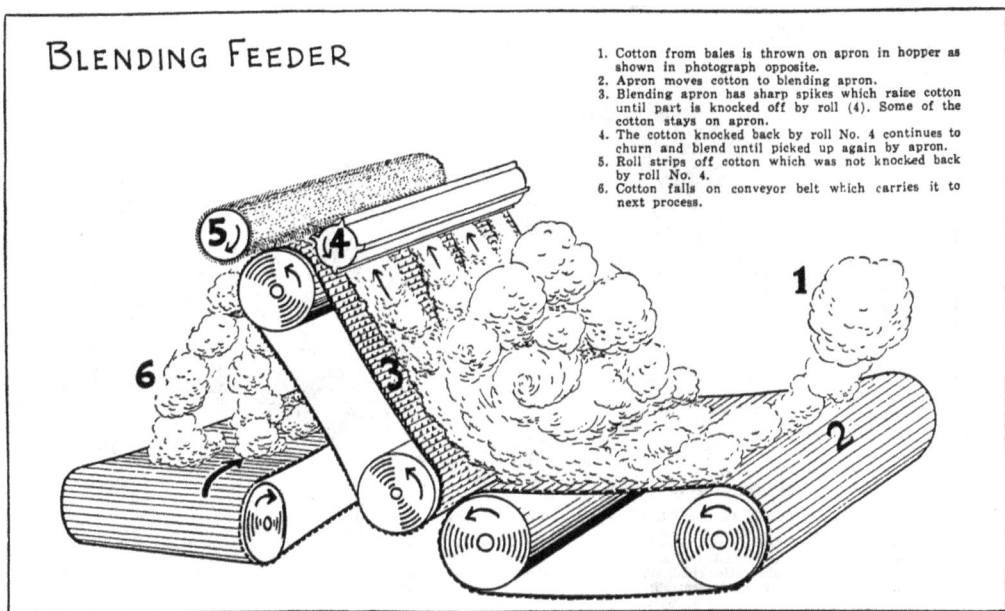

fig. 9.15b **How raw cotton is transformed into thread in a textile mill**

From the Bibb Manufacturing Co., Inc. Macon, Georgia, as shown in Linton: *Natural and Manmade Textile Fibers*

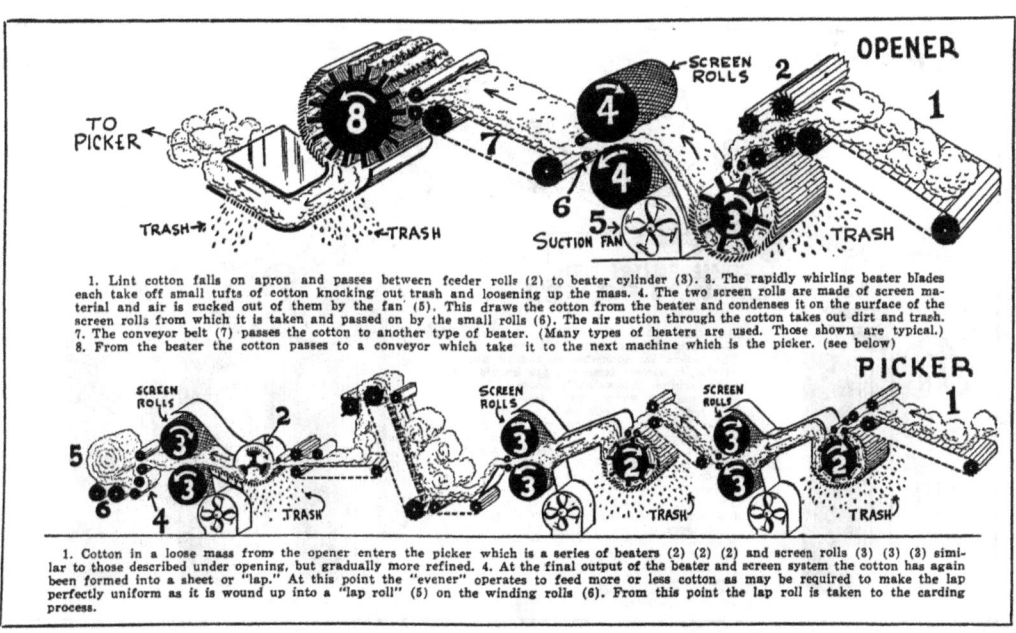

fig. 9.15c **How raw cotton is transformed into thread in a textile mill**

From the Bibb Manufacturing Co., Inc. Macon, Georgia, as shown in Linton: *Natural and Manmade Textile Fibers*

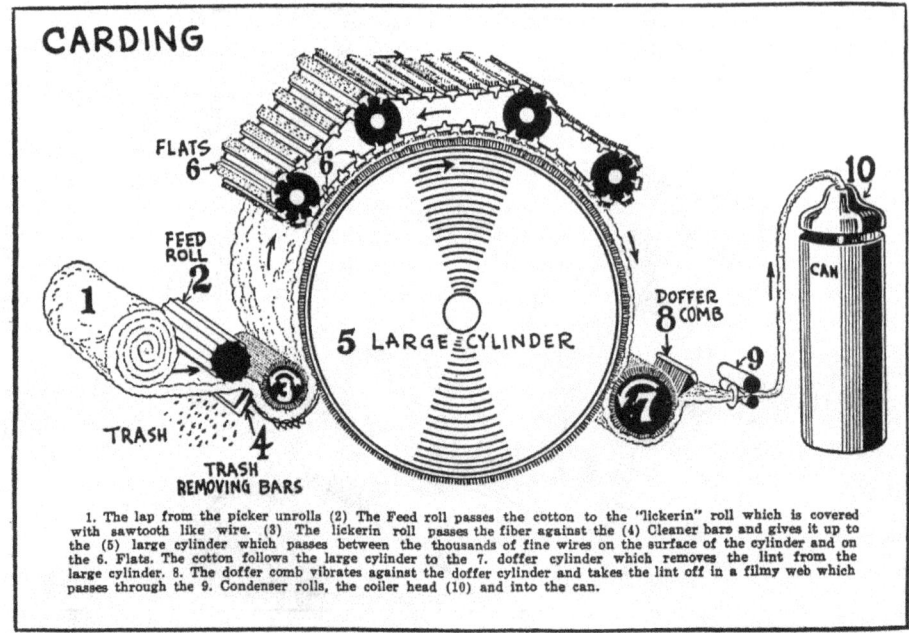

fig. 9.15d **How raw cotton is transformed into thread in a textile mill**

From the Bibb Manufacturing Co., Inc. Macon, Georgia, as shown in Linton:
Natural and Manmade Textile Fibers

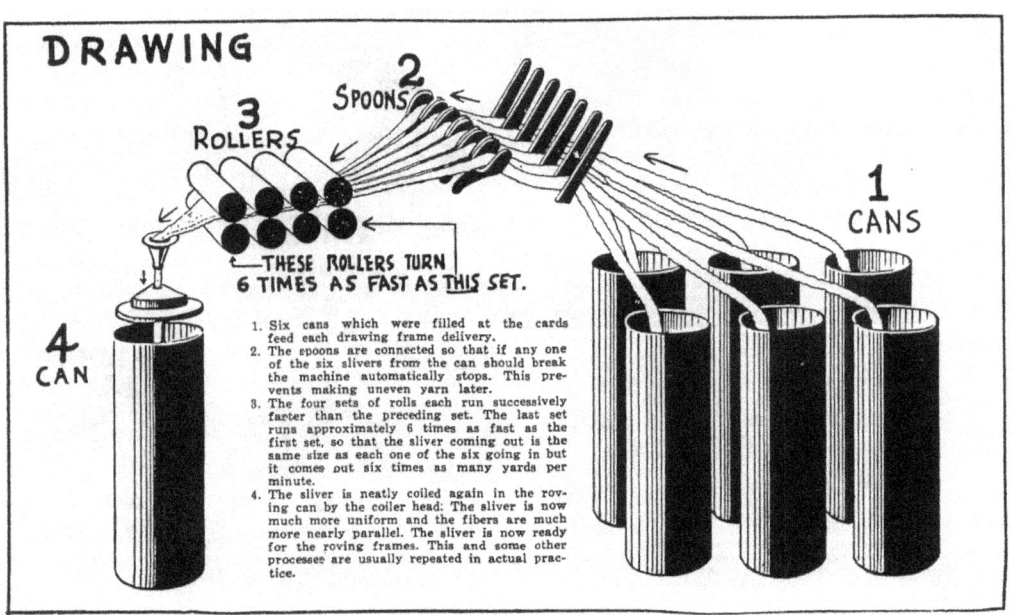

fig. 9.15e **How raw cotton is transformed into thread in a textile mill**

From the Bibb Manufacturing Co., Inc. Macon, Georgia, as shown in Linton:
Natural and Manmade Textile Fibers

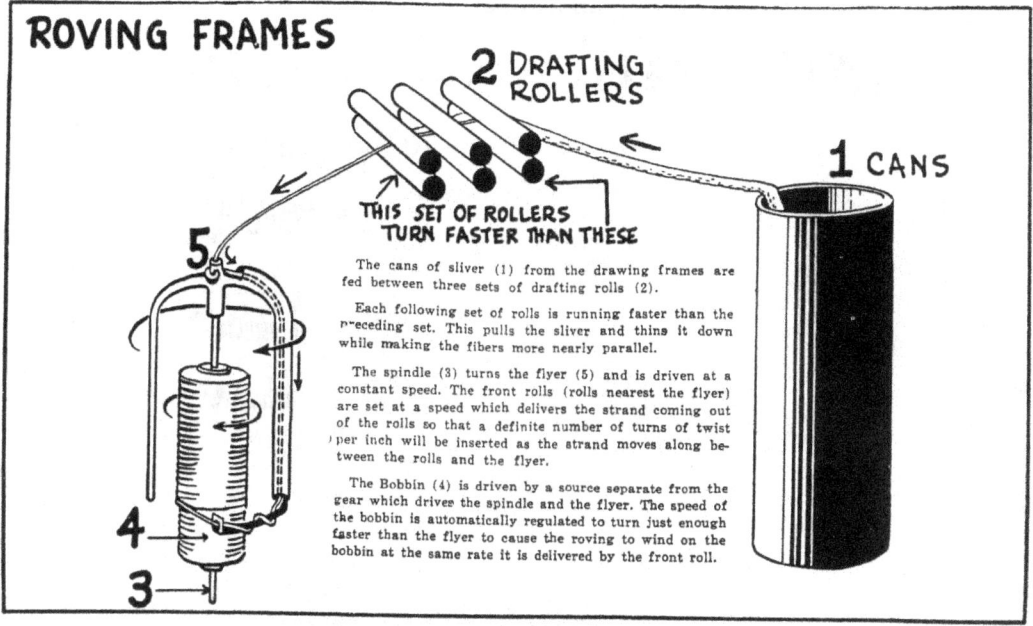

fig. 9.15f **How raw cotton is transformed into thread in a textile mill**

From the Bibb Manufacturing Co., Inc. Macon, Georgia, as shown in Linton: *Natural and Manmade Textile Fibers*

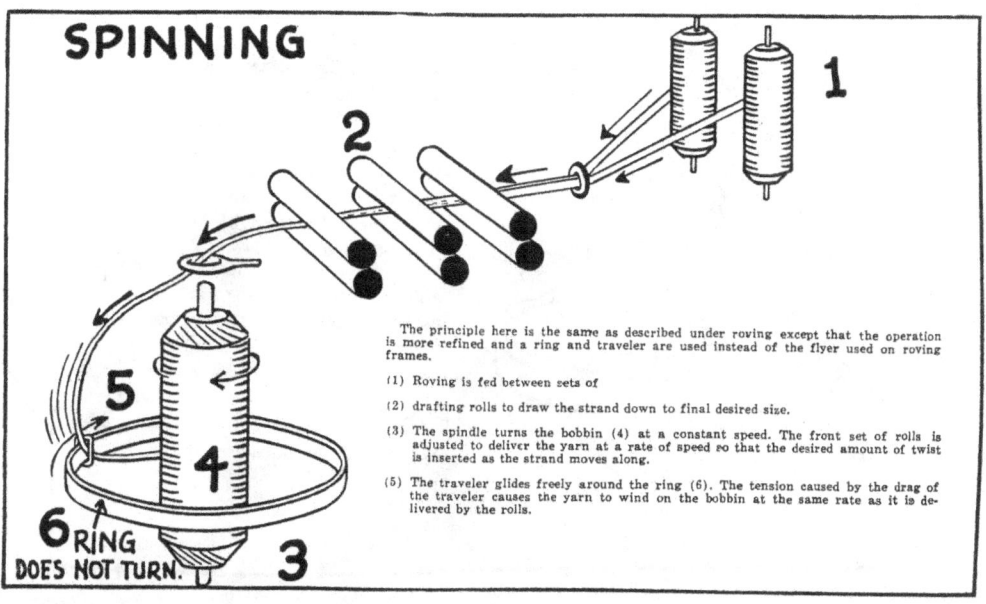

fig. 9.15g **How raw cotton is transformed into thread in a textile mill**

From the Bibb Manufacturing Co., Inc. Macon, Georgia, as shown in Linton: *Natural and Manmade Textile Fibers*

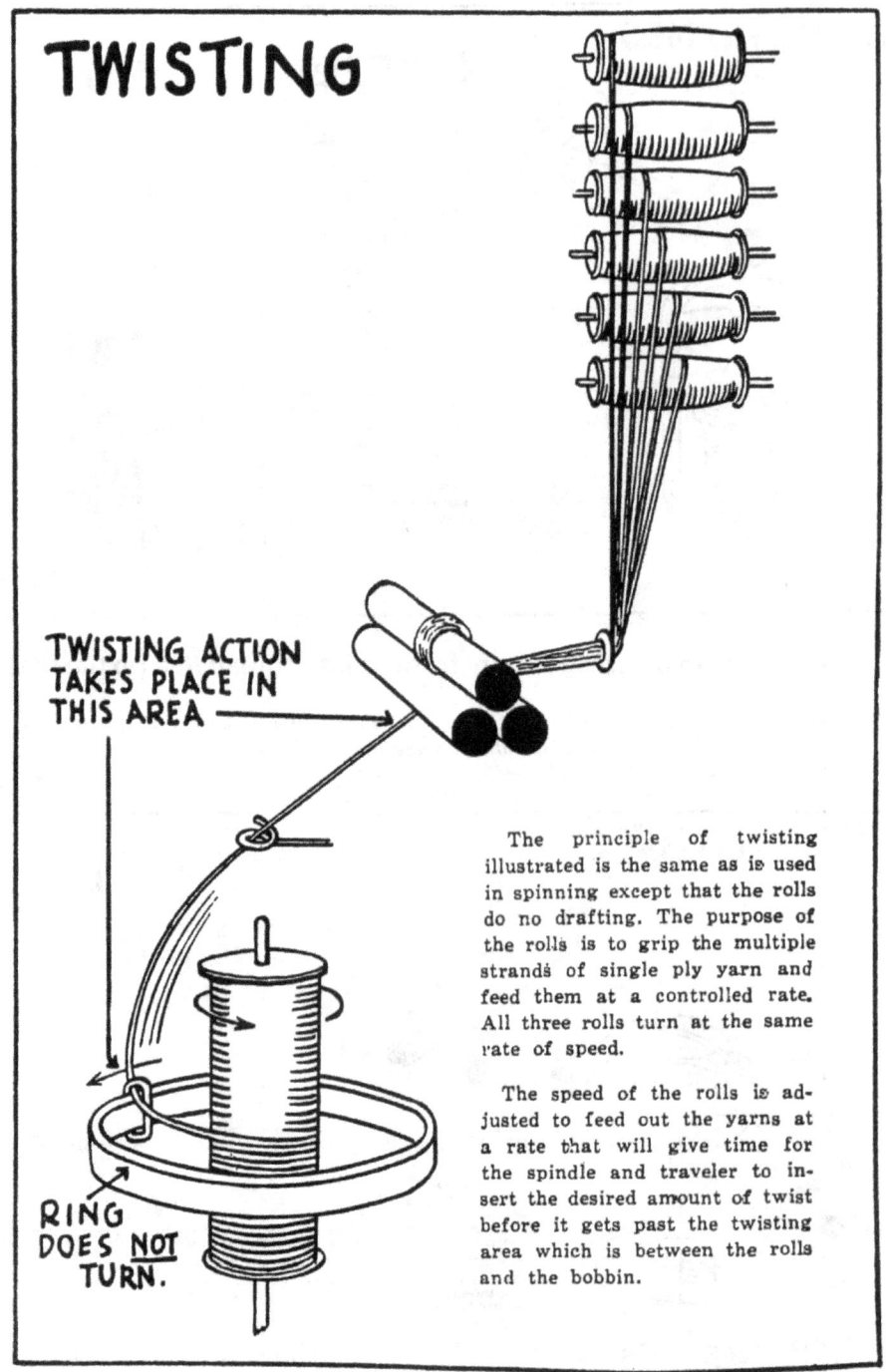

fig. 9.15h **How raw cotton is transformed into thread in a textile mill**

From the Bibb Manufacturing Co., Inc. Macon, Georgia, as shown in Linton: *Natural and Manmade Textile Fibers*

Breaking/picking: In the Fall River textile mills, the bales of cotton were received in the Picker Rooms, which were located in the rear ells of each mill. The bales were then opened and the clods of compressed cotton placed on a breaker-picker machine. The tumbling of the machine resulted in loose matter being removed and, with additional processing in the finisher-picker, the baled cotton was transformed into "lap" form, which were rolls of loose cotton about 40 inches wide weighing 40 pounds.

Carding: The cotton laps were then delivered to the Carding Room. The purpose of carding was to take raw cotton stock and make it into parallel sliver form. The carding process further removes debris from the cotton and disentangles the fibers, resulting in a loose rope-form of parallel cotton fibers.

Combing: This fine, detailed operation further cleanses the fibers and removes the short undesirable fibers, too short for combed yarns. Combed yarn is superior to carded yarn and is used in higher-quality fabrics and commands a higher price.

Drawing: The drawing process takes the loose thick cotton stock and reduces the diameter of the strand by stretching it out through the use of rollers. The individual fibers are not stretched at all; the diminishing of the strand diameter results from the attenuation as it progresses through the rollers.

Slubber and roving frames: This process further condenses the diameter of the strand until it reaches the diameter of lead in a lead pencil. Roving further draws out the fibers until they are ready to be spun into thread.

Spinning: Spinning is the final drawing and twisting of the cotton strand into a thread and the winding of the thread onto a practical device such as a cone or spindle. Spinning is done either on the ring frame or the mule frame.

Weaving: The thread or yarn is then brought to the looms to be woven into fabric.

Bleaching: Woven cotton fabric is produced in a "gray goods" state. It contains impurities and blemishes and is generally not appealing to the eye. As Linton notes, "There has probably never been a perfect yard of fabric woven on a loom. Fabrics are made in the finishing." (53-51) The first step in the finishing process is bleaching the fabric by its immersion in water solutions. Bleaching removes impurities and blemishes and makes the cloth white or very nearly white. It also aids in the affinity of dyestuffs and gives a better color effect.

Dyeing: If a fiber or fabric is to be a solid color, it undergoes dying.

Printing: Patterns or designs on cotton cloth are produced by printing. There are several methods of printing cotton cloth, including block, screen or stencil, or roller.

Most textile mills in Fall River took raw cotton and processed it into yarn and wove the yarn into gray goods fabric. Some mills manufactured only thread (Kerr Mills), some only bleached the gray goods into white fabric (the Fall River Bleachery) and some mills only printed the white cloth (American Printing Company). Occasionally, all operations, from raw cotton to printed textiles, were manufactured on the same site. This was true of the Chace Mill on Rodman Street, which, after 1895, added bleaching, dying, and finishing to its operations.

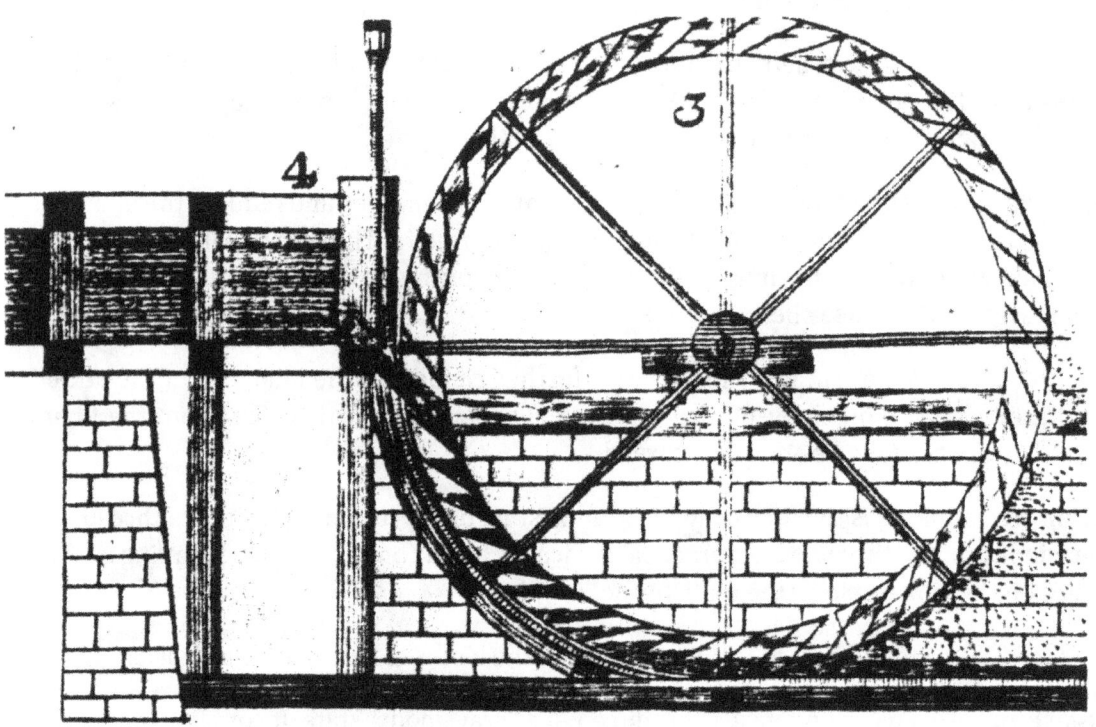

fig. 9.16 **A breast water wheel**

The breast water wheel was the most common wheel used during the early era of water power on the Quequechan River. Both kinetic and potential energy was used in this wheel type. If the wheel was designed so that the water hit below the wheel axis, it was called a "low breast" wheel; if it hit above the axis, it was called a "high breast" wheel. When water from the flume hit the wheel at the wheel axis, as shown in this sketch, it was called a "middling breast wheel."

From Oliver Evan's 1795 treatise: *The Young Millwright and Miller's Guide,* as reproduced in Dunwell: *The Run of the Mill*

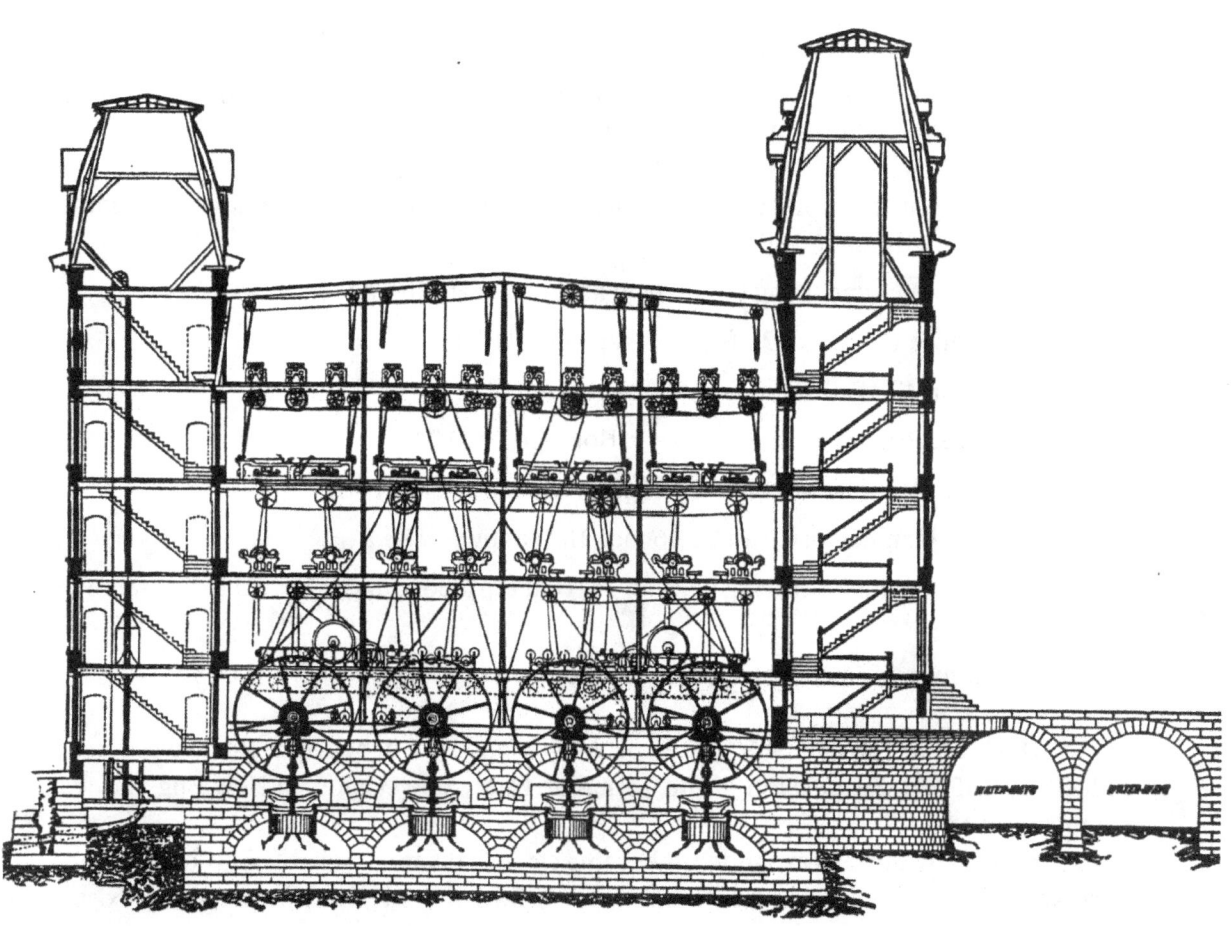

fig. 9.17 **Power transmission in a typical Rhode Island system textile mill**

This cross-section of the Manville Company No. 3 mill in Manville, Rhode Island, built about 1875, shows the transmission of power from the water trubine-type water wheels to the upper floor via belting systems. In this design—the state of the art for its time—each turbine had its own flume, tailrace, and a 2-foot diameter 10-ton flywheel. Dressing machines are on the second floor, cards on the third, mules on the fourth, and either looms or frames on the fifth, below the almost flat roof. The mill contained 400,000 square feet of floor area and was designed to use the full 1,600 horsepower of the Blackstone River on the 18-foot fall at this point in the river, a scale which the new power systems made possible. An 800 horsepower steam engine supplemented the turbines during the four to seven month period of low water.

From the 1875 catalog of the Leffel Company, which built the turbines depicted in the bottom of the mill.

Power and the Textile Industry in Fall River

Power was a central and all-important issue for Fall River's textile industry. Local textile machinery was powered first by hand, then by animal (Dexter Wheeler's horse-powered mill in Rehoboth), then in succession by waterpower, steam, and electricity. Each advance in power source provided greater productivity in textile production. In 1848, the following mills were located on the eight falls of the Quequechan:

- Troy Cotton and Woolen Manufactory: Fall 10.5 feet (depending upon the height of the pond), Horsepower 150.6.
- Pocasset Manufacturing Company: Fall 21.67 feet, Horsepower 225.4.
- Quequechan Mill: Fall 21 feet, Horsepower 218.4.
- Watuppa Mill: Fall 15.38 feet, Horsepower 160.
- Fall River Print Works: Fall 10 feet, Horsepower 104.
- Fall River Manufactory: Fall 14.46 feet, Horsepower 150.4.
- Anawan Manufactory: Fall 14.46 feet, Horsepower 153.2
- Metacomet Manufacturing Co.: Fall 16-18 feet(depending on the tide), Horsepower 148.2.

Total minimum fall 129.24 feet. Total Horsepower 1310.2

By 1850, all of the available water privileges on the Quequechan River were used. Indeed, by 1855, the expansion of the 15 major water power sites in New England was complete. (73-43)

As the mills on the falls expanded, it became clear that a more reliable flow of water was necessary. During the Quequechan's summer dry flow periods, insufficient water existed to power the mills, closing them down on a seasonal basis. Workers typically went back to their families' farms to work during the summer and early fall, took seasonal farm labor elsewhere, or remained seasonally unemployed.

However, this seasonal lack of power was damaging to the city's young textile and iron industries, and a method was sought to assure that an adequate flow of water was available during all months of the year. The logical way to accomplish this was to dam the river above the top of the falls so that the floodwaters would be held back during the wet seasons and gradually released during the dry seasons. However, state legislative approval was required for this action.

The result was the formation of the Watuppa Reservoir Company by a legislative act in 1826. The act allowed the Company to raise the level of the river two additional feet, with the Company paying all damages to property owners whose land would be flooded by the dam. This new dam flooded the river five feet above its original level, causing extensive

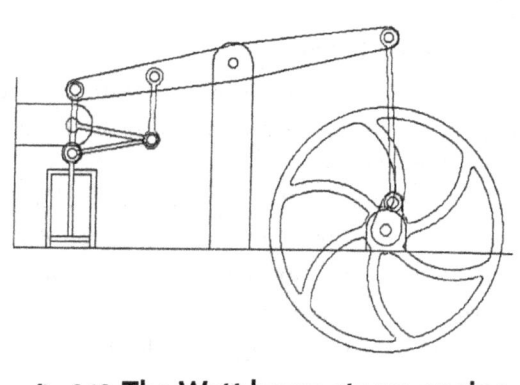

fig. 9.18 **The Watt beam steam engine**

The Watt beam steam engine

One of Watt's major contributions to the Industrial Revolution was that he adapted the steam engine by adding a crank and flywheel that could power machines, thereby replacing water wheels. The 1763 beam engine shown here, as modified later, powered factories and steamboats. While an improvement over the Newcomen steam engine, the Watt engine was still inefficient, used substantial amounts of fuel and was not uniform in motion.

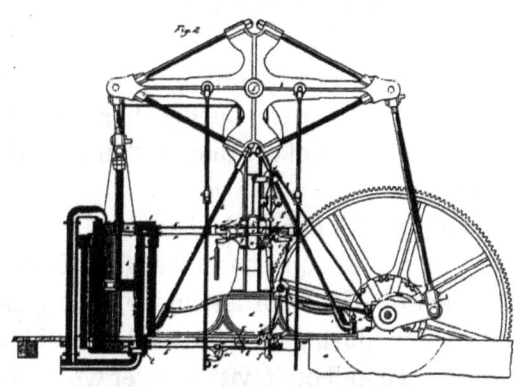

fig. 9.19 **The Corliss steam engine**

The Corliss steam engine

George H. Corliss of Providence received 67 patents related to improvements in the steam engine. He improved the valve gear regulated by the centrifugal governor, which revolutionized steam engine design by providing for uniform motion regardless of load and by reducing the waste of steam, thereby dramatically reducing fuel costs. If all but one of 100 looms suddenly stopped in a mill, the Corliss engine would allow that one loom to continue working at the same rate of speed

flooding of the marshes and flat land above the dam. The dam caused the flooding of the area that is now Ruggles Park and Stafford Square, among other flat areas along the river that were eventually filled in for development.

The new dam created a storage capacity of 250 million gallons of water, assuring a more even flow of water for the mills during any season. The Reservoir Company operated the gates of the dam and opened them when more water was needed.

Water-powered textile mills then faced the challenge of finding ways to increase the efficiency of their fixed amount of water power. The earliest wheels were of the high breast design, where water enters the wheel above the axis; however, these wheels never harnessed more than 60 percent of the potential available water power. Machines had to be kept relatively close to the main shaft of the waterwheel, otherwise the wooden shafts transferring power to the machines absorbed the power in flexing and twisting.

Later, progress in mathematics and hydraulic analysis resulted in the invention of the turbine wheel by Benoit Fourneyron in France in 1832. The turbine wheel incorporated precisely curved vanes inside the turbine, resulting in increases of efficiency to 75 percent. (24-62)

Gradual improvements in turbine design by engineers in Lowell resulted in efficiencies of up to 88 percent. Conversion to turbine water-powered wheels, coupled with improvements in metallurgy, allowed textile mills on the Quequechan and elsewhere to increase their machinery and enlarge their capacity.

In Fall River, the Kilburn Lincoln Company was formed in 1845 by several mechanics who began manufacturing the Kilburn Water Turbine. The company declared itself ready to manufacture a wide variety of textile manufacturing and iron making machinery, but, over time, it specialized in cotton machinery, particularly high-speed plain looms. (73-52)

In the first Massasoit Mill (later renamed the Watuppa) on the falls, built in 1827, Holder Borden first introduced leather belts for the transmission of power, replacing gears "by which much of the noise and racket of machinery was done away with, and a steady and more uniform motion secured to the different processes, to say nothing of the reduction of friction and gain in power." (25-24)

In 1847, the first water-powered cotton mill in Fall River incorporating a steam engine, the Metacomet Steam Cotton Mill, was built on the lowest falls on the Quequechan. In 1848, the Massasoit Steam Mills, built on Davol Street near the mouth of the Quequechan, was the first factory in the city powered solely with steam power.

As with most inventions, the new steam process was not adopted all at once, but was used to supplement water power in existing mills on the Quequechan River falls. During its early years of use, particularly before the introduction of the Corliss engine, steam was considered as a backup to water power, especially useful during droughts and annual low water periods. The Metacomet Mill is a good example of a mill built with a water wheel but which also incorporated steam.

However, the early steam engines had the disadvantage of having unreliable speed. With the introduction in 1848 of an improved steam engine invented by George Corliss of Providence—providing an engine that could control speed—steam power was revolutionized.

With the invention of the reliable Corliss steam engine, steam power released mills in Fall River and elsewhere from the reliance on water power sites. The first Corliss engine in Fall River was introduced at Union Mill No. 1 in 1859, and from that time the use of steam power increased rapidly in the city. Further refinements in steam power, particularly the invention of the compound-condensing engine, allowed mills to increase their capacity exponentially. While the largest mill using water power on the Quequechan River had an output of slightly more than 200 horsepower, refined steam power in the modern factories provided up to 3,000 horsepower.

With the introduction of steam, the Quequechan River assumed a new role. While it continued to provide water power for mills on the falls for many more years, the principal function of the Quequechan now became one of providing water for making steam and for cooling and condensing the steam once it passed through the engines.

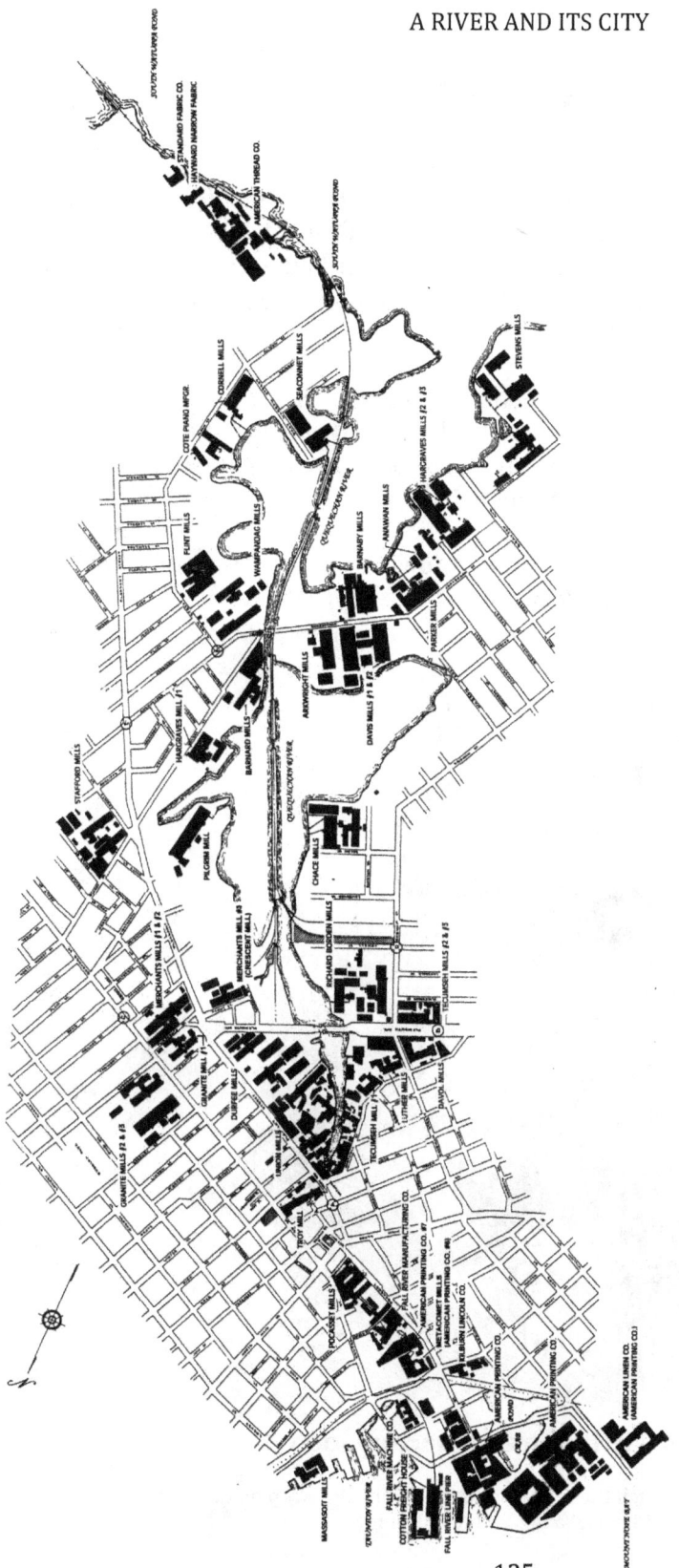

fig. 9.20 **Textile mills located on the Quequechan River, 1915**

Base map from Faye Spofford and Thorndike: 1915 study of Quequechan remediation, with additional data from Sanborn maps of 1900.

Free of the constraint of having to be sited on the falls of the Quequechan, new steam-powered mills began being built along the Taunton River, on Mount Hope Bay, on Cook Pond and Laurel Brook, and on the upper flat section of the Quequechan River, financed by the profits from the early water-powered mills. Between 1855 and 1865, the number of spindles in the city doubled and between 1865 and 1875 quadrupled. In the late 1860s, Fall River surpassed Lowell as the largest textile-producing center in the United States. (73-54)

The development of the Corliss engine and of reliable steam power gave Fall River a major competitive advantage:

> While water power had led to a dispersion of isolated factories, steam mill economy favored concentration at a few coastal locations where transportation costs were minimized, labor was plentiful, and local investment capital was available. As the Niles Weekly Register put it: 'It is cheaper to use steam power in the midst of a dense population, than to use water power, which often makes it necessary not only to build a factory, but a town also.' New England's port cities, especially Fall River and New Bedford, offered optimal conditions and hosted the region's most aggressive steam mill development. (24-105)

fig. 9.21 **Mill girl at an early factory loom**

A mill girl draws warp yarns through the reed prior to mounting the warp on the loom, 1845.
Merrimack Valley Textile Museum

The Maturing Textile Industry in Fall River

Fall River took advantage of the prosperity that followed the Civil War. "It had capital, skilled labor, and courageous management. Its coastal location favored steam power and gave print cloth manufacturers a small but significant edge over competitors along the rivers to the north." (24-106)

From 1863 to 1868, eight new companies brought their mills into production, more than doubling the city's capacity to over one-half million spindles. After 1870, an enormous surge of mill building occurred. Between February 1871 and March,1872—a period of only 13 months—15 new textile corporations were formed and 22 mills were built, an amazing feat for such a short time period. (73-51) As Dunwell describes in *The Run of the Mill*:

> By 1876, Fall River's forty-three factories had an installed capacity of well over one million spindles, feeding yarn to more than 30,000 looms. Fall River grew faster than any other textile city of its time. One out of four new spindles added to New England between 1850 and 1875 was placed in Fall River. The city controlled one-sixth of all New England cotton capacity, and one-half of all print cloth production. Fall River rightly called itself 'Spindle City'—preeminent in America, second only to Manchester, England, in the world. (24-106)

Fall River was a thoroughly one-industry town, with over 14,000 of the city's 16,000 wage earners employed in the cotton mills. Not only was Fall River solely a cotton textile manufacturing town, but within that specialization it specialized in cotton print cloth to the exclusion of all others. Of the 33 corporations in the city, 26 listed print cloth as their only product.

However, the "gray" cloth produced from raw cotton, if it was to be marketable, had to be bleached and printed. The city's bleaching capacity was increased substantially in 1872, when the Fall River Bleachery was built on Sucker Brook, a source of pure process water that flowed out of Stafford Pond in Tiverton and into the South Watuppa Pond.

The finishing and printing industry had started in the 1820s and had grown significantly as the production of cloth increased. Fall River's printing capacity kept up with cloth production until about 1855, when cloth production and calico printing balanced out at 19,000,000 yards annually. The Iron Works group started the American Print Works, later the American Printing Company, and gradually increased its capacity until it became the largest print works in the United States, with an annual capacity of 80,000,000 yards of cloth. Together with the Bay State Print Works, located in the old Globe Mills and which came under the control of the American Printing Company, the printing capacity of the city increased to 100,000,000 yards a year. (73-51 and 52)

However, this was only a third of Fall River's annual production of 300 million yards of print cloth. The remaining two-thirds of cloth production was sent to finishing and printing mills located principally around New York and Philadelphia. (73-51, 52 and 66)

fig. 9.22 **Mule spinning**

A typical English spinning mule. Turning the large hand wheel (center) moves the carriage back and forth on its rails. The mule was capable of subtle, delicate spinning, and it was ideally suited for wool and fine cotton filling, or weft, yarns where tensile strength was not required. Like the jenny, it could not be used for the warp. A contemporary observer noted that a typical mule spinner walked over 25 miles a day.

Source: Baines: *Cotton Manufacture*

By 1910, the full growth of Fall River's textile industry had been reached, with 43 corporations, 222 mills, and 3,800,000 spindles. (14 -9) Fall River's textile mills produced two miles of cloth in every minute of every working day in the year. (88-3)

The Importance of Textile Machinery

Blacksmiths played a central role in the development of machinery that propelled Fall River into the Industrial Revolution. Dexter Wheeler was a blacksmith from Swansea who built the first power cotton spinning equipment to be used on a mill on the Quequechan River—the Fall River Manufactory—and later built the first power looms to be used in Fall River in that same mill. (25-13; 25-16)

John Borden Jr. and his brothers learned blacksmithing in their father's shop and went to the factory of Francis Cabot Lowell in Waltham to learn how to make textile manufacturing machinery. When they returned to Fall River, they manufactured the spinning machinery installed in 1813 in the Troy Cotton and Woolen Manufactory, the second textile mill to begin operations on the Quequechan. The Troy mill included space for a machine shop and a blacksmith shop, with two forges, which was rented to John Borden Jr. (25-15)

The beginnings of the Fall River Iron Works, later to become the American Printing Company, began when Colonel Richard Borden and Major Bradford Durfee experimented in the local blacksmith shop, manufacturing spikes, bars, rods, and other articles of iron. (25-44)

Blacksmiths were the first machinists, and machinist shops were set up in each early textile mill in Fall River so that the machinery could be built as the building was being raised. (24-53) Andrew Robeson's Fall River Print Works on the upper Quequechan had its own blacksmith shop that constructed what is considered to be the first textile printing machine made in the United States. (25-30)

As the demand for textile machines increased, blacksmiths established independent machinist firms to supply the demand. The first of the separate machine-building companies, Harris, Hawes, and Co., was formed in 1821 and occupied two floors of one of the Pocasset Company mills. The Harris, Hawes firm made most of the machinery for the Pocasset or Bridge Mill, and most of the improvements in the machinery of the Fall River Manufactory and the Troy mills were made by this firm. (25-26)

From the very beginning, textile manufacturing in factories required machinery of ever-growing complexity. The building of textile machines is a specialized art requiring designs, skills, and the proper tools and materials. The first textile machines in America were built entirely of wood. Embargoes placed on the export of machinery from the British Isles retarded the growth of the young textile industry in America. Without adequate machine shops, each early mill had to provide its own. As soon as the basement foundation of a mill was completed, a machine shop began operating there to provide the machines that would be needed to equip the floors above. (24-53)

In the early years of textile mill development, blacksmith/mechanics set up their own shops and produced a great variety of machines necessary for mill construction and textile production. During these formative years, specialization was impossible. However, as the industry grew, textile machine production became more sophisticated and specialized, with some shops specializing in certain types of textile machinery. (73-52)

Not until the 1820s did machine tools reach a level of sophistication where they could begin reproducing themselves. The early wooden machines evolved into metal machines made of brass, wrought iron, and cast iron, each having its specialized uses on the machines that fit the characteristics of the metal. However, it wasn't until the 1850s that versatile steel was introduced. Early mill owners hired the best mechanics to convert their designs into wood and metal machines.

By their very nature, machines evolve towards increased efficiency, automation and versatility. Squeezed between increasing competition and rising labor costs, manufacturers encouraged mill machinists to create machines that were bigger, faster, and simpler to operate. Invention served the manufacturer by minimizing his dependence on expensive skilled laborers and maximizing productivity of the unskilled operatives who took their places. An ideally mechanized mill would be staffed entirely by unskilled labor tending as many machines as humanly possible, running at maximum theoretical speed. (24-54)

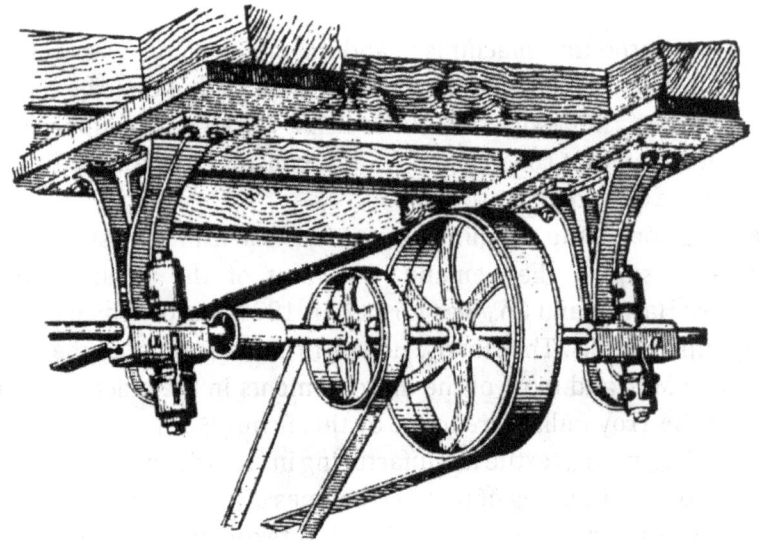

fig. 9.23 **A countershaft with self-aligning bearings**

Shafts, pulleys, and bearings had to be carefully machined to turn perfectly at speeds that, by 1860, often exceeded 300 rpm.

Source: Knight: *Mechanical Dictionary*

Mule spinning required a strong, skilled male operator to move its carriage back and forth. Because the skill was very difficult to perfect, mill owners relied on trained English mule spinners who also happened to be notoriously independent and brought with them the craft union tradition and hostility towards management. From the perspective of the mill owners, mule spinners were "the most troublesome class of operators in the mill." Mule spinners proved that "the more skillful the workman, the more self-willed and intractable he is apt to become." (24-16)

In their zeal to rid themselves of the independent English mule spinners, the Fall River mill owners were intent on acquiring the self-acting mule, recently invented and in use in England. In 1838, William C. Davol and Bradford Durfee visited Manchester, England, to purchase some of the new Sharp and Roberts self-acting mules under an agreement with Sharp where they would manufacture the mules for Sharp under an American patent. However, England had restrictions on the removal of new industrial technology from the country.

Undeterred, and in an audacious act of industrial espionage, Davol and Durfee arranged with agents in England to have the machinery sawn into small pieces, packed into a thin case marked plate glass, sent to France, then on to Fall River, where it arrived two years after its purchase. Once in Fall River, the pieces were reassembled into a model and then into many machines in Davol's machine shop, Hawes, Marvel, and Davol. However, it was not until 1846 that they installed the mules in the new Metacomet Mill, which Davol and Durfee built from plans brought from England. Fall River was the first textile center in America to use Sharp and Roberts' self-acting mules. (81-228)

However, the plans of the mill owners were thwarted, both because of the organized resistance of the mule spinners and because the new machinery was so complicated that

fig. 9.24 **The Fall River Water Works Power Station on North Watuppa Pond**

This is one of the many municipal buildings of magnificent design and built of Fall River granite. Unfortunately, it is deteriorating due to many years of neglect.

it still needed the skills of the mule spinners to operate it. Mule spinners continued their dominance for several more decades. (24-56)

Calico printing machines imported from England also required experienced English workman to operate the demanding process, but the printers were as independent as the mule spinners.

One of the major challenges of textile mills from the very beginning was how to transfer power from the water wheel to the upper floors of the building and then to the individual spinning machines and looms. Before the development of metalworking in America, early mills used wood for the water wheels, the gears, and the shafting. However, this method of transferring power in the mill from the water wheel to the textile machines was clumsy and inefficient. When wrought iron shafts replaced their wooden predecessors, they were heavy and rarely uniform in strength. Instances were common where mill floors collapsed under the weight of their shafts. The problems caused by these early shafting systems were magnified as speed increased, with the result that advances in speed of textile machines were retarded by the limitations of transferring power from the wheel to the upper floors.

In 1826, Paul Moody developed a system in Lowell of leather belts that carried power from the wheel to intermediate countershafts on each floor and from the countershafts to the machines. This belting system was further perfected by Zachariah Allen at his mill at Allendale, Rhode Island, in the 1830s. In Fall River, the enterprising Holder Borden leased the Massasoit Mill in 1831 and modified it so that it incorporated a belting system, the first in the area. (25-24)

In 1836, a Scotsman named James Montgomery visited America and, following an inspection of the Massasoit and another mill in the town, described the clear advantages of belting systems:

> There are two mills at Fall River, in the State of Rhode Island, which seem to decide the question in favor of belts. These factories have equal water-power, as the one takes exactly what passes through the other. The one is geared with belts, the other with shafts, etc., and it is found that the other can put in motion considerably greater quantity of machinery than the latter. (25-25)

Advances in the cold rolling of steel in the 1840s allowed the production of strong, light shafts, and improved lathes made possible more precise bearings, pulleys, and flywheels. These advances in power transmission—coupled with the development of high-speed turbine water wheels—resulted in the universal adoption of the belting system in the Fall River mills beginning in the 1840s. This development led, in turn, to the emergence of a local leather industry created to provide the belting systems for the Fall River mills.

Fall River was also the location for a thriving textile machinery business. Fenner's 1911 *History of Fall River* mentions that the Kilburn, Lincoln, and Company plant was one of the largest makers of looms for cotton and silk weaving in the United States, making about 5,000 looms annually at that time. The plant also made power transmission machinery.

Textile machinery manufacturing in the city first began prior to 1840, when Jonathan Thayer Lincoln began the business.

The perfection and upgrading of machinery allowed Fall River to produce finer goods and, therefore, compete more successfully with Southern manufacturers than other Northern textile centers. As Henry M. Fenner says in his 1906 edition of the *History of Fall River*:

> The lesson of the necessity of the best of modern machinery, of liberal allowance for depreciation and of competent management, has been learned and will not be forgotten. The improvement in machinery has been so rapid that present equipment is no longer allowed to wear out, and is discarded to make way for new to meet competition. (31-74)

Riparian Rights to the Quequechan River

Shortly after the beginning of the textile manufacturing on the Quequechan River, a perennial issue was how to assure an adequate flow of water during all seasons and all years. Seasonal variations and occasional droughts were taking a toll on the viability of the industries on the river. Orders for gray cloth, printed cloth, or iron products could not be delivered with reliability if the flow of the river was not predictable. However, no one interest in the river could resolve the issue; a common solution was necessary.

In 1714, through the purchase of shares owned by Colonel Benjamin Church, Richard and Thomas Borden, the sons of John Borden, became sole owners of the south side of the Quequechan River. Since John Borden had previously purchased the land and water power north of the falls, in that year the Borden family came into sole possession of the entire water power on the Quequechan River. (26a-424)

In 1813, in exchange for shares in The Troy Company, the heir to John Borden, Mrs. Amey Borden, conceded control of the upper falls to the company, where its building was located. This concession resulted in the Troy Company in effect controlling all of the water power on the river. (25-25) This control of the water privilege by the Troy Company was monitored closely and "any violation or invasion of its rights was jealously watched and guarded against." (25-26)

As more textile mills began to be built on the river, however, "it became necessary to establish a general and responsible control of the water-power furnished by the stream and the parent lake." (25-25) In 1822, representatives of the Troy and Pocasset mills conferred "upon a permanent mark for the height of the flowage of the pond." In 1825, following a general conference of all the affected parties, the question of the permanent preservation and control of the river's water power was settled. The result was the formation of the Watuppa Reservoir Company in 1826, which was officially created by the legislatures of both Massachusetts and Rhode Island, since the watershed of the river—and the river itself at that time—was in both states. The Watuppa Reservoir Company was formed "to build a new dam above the dam belonging to the Troy Company, for the purpose of raising the water two feet above the present dam, and to pay the expense of flowage occasioned thereby." (25-26)

The Reservoir Company proceeded to build the dam in 1832, paying for damages resulting from the rise in the water level above the dam. The dam was constructed of mortared quarried stone and its building supervised by Major Bradford Durfee.

The Company then turned its attention to controlling all of the flow of the ponds and streams that flowed into the Quequechan River, not only in Fall River but in Westport and Tiverton, Rhode Island. The directors of the Watuppa Reservoir Company were master planners and were determined to become absolute arbiters of all water flowing into the Quequechan River. In 1843, the Company ordered its agent "to use his best endeavors to prevent the water being turned, or any part of it, from any of the ponds that empty themselves into the one from which we draw our water, and for him to pay our proportion of all expenses that may arise from legal or other means that shall be deemed proper to prevent the course of said waters being turned, either by digging, building, or otherwise." (25-22)

Through its purchase of the Troy Cotton and Woolen Manufactory and the Fall River Manufactory, among other interests, the Fall River Iron Works soon became the controlling interest in the Watuppa Reservoir Company. Colonel Richard Borden was its prime mover and its president. This control allowed the Iron Works to maintain a reliable flow of water to its iron and textile operations on the falls.

Up until the 1870s, the water flowing down the Quequechan River falls was used commonly and solely among industrial users. However, a new user soon demanded its share of the water from the river's watershed. This new user did not want to use the water of the Quequechan River but instead wanted to actually reduce the flow of the river by taking away a major portion of the watershed, a new and alarming prospect not anticipated when the Watuppa Reservoir Company was formed.

As a result of the rapid mill development that occurred in the 1870s, the city's population exploded with new immigrants. It therefore became imperative that a clean source of public water be made available for residents, including a modern system of distribution. The makeshift domestic wells in the city were often polluted by nearby cesspools and outhouses, and the frequent outbreaks of cholera demanded that corrective action finally be taken.

The only practical source for a public water supply was the North Watuppa Pond. However, taking North Watuppa's water out of the watershed of the Quequechan River meant that there would be less water for use by the city's textile mills. In a series of steps, beginning in 1871 and ending in 1892 with an agreement with the Watuppa Reservoir Company, state legislation allowed the City of Fall River limited rights to the use of the water of North Watuppa Pond.

The agreement and legislation stipulated that when the South Watuppa Pond fell below 40 inches below full pond, the city must allow North Watuppa Pond to flow into the South

Watuppa and that, under such conditions, North Watuppa could not be kept more than one inch above the South Watuppa Pond. In effect, when the South Watuppa Pond fell 40 inches below full pond, the city would lose all control over its water supply. (37-6)

In practice, the city allows no water to flow into the South Watuppa Pond except during periods of water overflow. The city acquired the rights to the water of the Noquochoke River and pond in Westport and Dartmouth and constructed a pumping station and a transmission line to the South Watuppa Pond. The city pumps Noquochoke water to the South Watuppa Pond during the wet season to keep the South Watuppa as high as possible during dry periods. This action virtually eliminated the need for the Watuppa Reservoir Company to tap into the North Watuppa Pond. (43-8)

As the principal owners of the Watuppa Reservoir Company, the American Printing Company (successor to Richard Borden's Fall River Iron Works) stipulated in the 1892 agreement that the Printing Company be guaranteed a minimum amount of flow even during the most severe droughts. When the water level fell below 40 inches below full pond, the Reservoir Company could not allow more than five million gallons a day through the Third Street dam and not less than three million gallons a day if the American Printing Company required it. If three million gallons would not flow, the Reservoir Company would have to limit the flow to allow the Printing Company to have at least two million gallons a day from the Quequechan. If two million gallons of water a day could not be delivered to the Printing Company by the Reservoir Company, then the city would be required to make up the difference through its own supply. (43-7)

On January 9, 1919, the Watuppa Reservoir Company relinquished all of its rights to the North Watuppa Pond and conveyed them to the City of Fall River.

10 The Architecture and Evolution of Fall River's Textile Mills

The Four Periods of Mill Development in Fall River

One of the unique aspects of the Quequechan River is its textile mills and their architectural evolution. There were four distinct periods of mill building in the city: (1) the earliest mills built between 1813 and 1839 on the Quequechan falls; (2) mills built between 1840 and 1858 on the Quequechan falls; (3) mills built from 1859 to 1879 away from the falls, on flat land above and below the falls; and (4) the wider flat-roofed mills built from 1880 to 1910. These major periods of mill building reflect the economic cycles and technological innovations of the textile industry.

The first mills

The first era of textile mills in Fall River were the small water-powered structures built during the earliest years of the city's budding textile industry. These first mills were built over the Quequechan River falls and were narrow in width, given the topographic restrictions of building on a steep hillside and the need to allow natural light to penetrate into the interior of the mill building. They also had small windows, probably a result of their masonry and wood construction, which limited the length of open spans. Most of these mills were built of fieldstone; however, other materials were used, such as wood, brick, and, occasionally, granite.

The first cotton textile mill in Fall River was Colonel Joseph Durfee's mill, built in 1811, at what is now Father Kelly Park at Globe Corners. This first mill was a one-story wooden structure, with its water power fed from a dammed stream from Cook Pond that formed a pond where the park is now located.

The first textile mills on the Quequechan River were the Fall River Manufactory, completed in October 1813, and the Troy Cotton and Woolen Manufactory, completed in September, 1813 (but did not commence operations until March, 1814). The Fall River Manufactory building measured 60 feet by 40 feet and was three stories in height. The first story was built of fieldstone and the upper stories built of wood because, as was said at the time, "there was not enough stone in Fall River to finish it with." (25-13) However, that is unlikely, given the amount of fieldstone that was available in the immediate area. Another reason given is that stone was brought up to the masons on ramps and, for the upper floors, this was impractical, since derricks were not in use in the area at that time. (25-13)

The Troy Mill measured 108 feet by 37 feet, was four stories and was built entirely of fieldstone from neighboring farm fields. However, it remains unexplained why the Troy Mill could be built entirely of fieldstone but not the Fall River Manufactory structure.

With the relaxation of import tariffs following the War of 1812, and the resulting flood of cheap English textiles into America, the young New England textile industry collapsed. As a consequence, no new mill construction occurred on the Quequechan River for several years.

Beginning in 1821, following the restoration of tariffs, the Quequechan experienced a minor building boom that continued for 18 years. Samuel Rodman of New Bedford built the first Pocasset Mill in 1821, also known as the "Bridge Mill," situated just west of Main Street. This three-story mill was constructed of stone and measured 100 feet by 40 feet, with a long ell that extended over the "stream." It accommodated 1,000 spindles in one-half of the building; the remainder was leased to another textile manufacturing firm. (25-23) The first print cloths made in the village were woven in the Bridge Mill and printed in Andrew Robeson's print works. (25-28)

Samuel Rodman's Pocasset Company tended to specialize in constructing buildings for lease to various small firms. One of these buildings was the Satinet Factory, erected in 1825, a part of which was leased to Andrew Robeson for his calico printing business. The other half of the building was occupied by J. & J. Eddy for the manufacture of woolen textiles. In 1826, the "New Pocasset" was erected; it was built of stone and leased for making cotton yarn and cloth. (25-23) "Thus had the Pocasset Company fostered the manufacturing enterprise of those days by providing a place to make beginnings." (25-27)

In 1825, the Anawan Manufactory was organized and a mill built on the Quequechan

River opposite where Anawan and Pocasset Streets merge. The Anawan Mill, built by Major Bradford Durfee, was an architectural hybrid, for it had hammered granite walls for the lower two stories and brick for the upper stories that was fired at Bowenville from clay brought from Long Island. The mill was large for its time, running between 5,000 and 7,000 spindles. Providence and New Bedford capital funded the mill. The names from Providence included Wilkinson, Valentine, and Butler. The New Bedford investors included the whaling families of Rodman, Rotch, Swain, and Morgan (Charles W. Morgan). (25-27)

Andrew Robeson's print works soon outgrew the Satinet factory and in 1826 he built new buildings for the successful print works later known as the Fall River Print Works. As the print business prospered, Robeson added new buildings until the last and largest was added in 1836. Robeson once worked at his father's flour mills in Germantown, Pennsylvania, and he hired workmen from Pennsylvania to build his first mills in Fall River. They had an exterior coat of blue mortar, a novelty in this section of the country. (25-30)

In 1827, the largest mill yet constructed in the village was realized: the Massasoit Mill, later known as the Watuppa Mill. It was so large that it was believed that no one firm would consider leasing the whole building, so a stone partition wall was erected from the foundation to the roof and two wheel pits put in. Holder Borden began managing this mill at the age of 31; he removed the partition and filled the building with machinery for making various fabrics. It was the first mill in the vicinity where belting was used to transfer power from a water wheel to the machinery above. (25-24)

The Fall River Manufactury was enlarged in 1827 with the addition of a small brick mill three stories in height. This was called the Nankeen Mill because it produced nankeen cloth. In 1839, both the original Fall River Manufactory mill building and the Nankeen Mill were torn down to make way for a new Fall River Manufactory building called the White Mill. (25-22)

At the last falls before tidewater, the Fall River Iron Works began building their furnace, foundries, and rolling and slitting mills after 1821. Not much is known about the construction of these buildings. They were all demolished to make way for the Metacomet Mill in 1847, when the Iron Works moved its operations to the wharves on Mount Hope Bay.

These early textile mill buildings reflected the vernacular architecture of wooden or stone English country barns, which at that time could be found on virtually every New England farm. These barns were rectangular with hipped roofs and small windows. The earliest mills also resembled the saw, grist, and fulling mills on the Quequechan River.

None of these early mills remain today because they occupied valuable real estate on the falls and were subsequently torn down to make way for larger, more modern structures.

The second phase of mill construction

The second period of textile mill construction in Fall River occurred in the 1840s, when a new generation of larger mills were built on the Quequechan River falls. These more substantial mills were built of rubble granite with parged or roughly cemented joints. The

fig 10.1 **The Metacomet Mill**

This is the oldest existing textile mill in the city and the only one remaining of the original water-powered mills on the Quequechan River. It was built on the lowest falls on the river at tidewater.

Following a visit to England to inspect a textile mill in Bolton that was reputed to be a model of English engineering, Major Bradford Durfee and William C. Davol collaborated with Jefferson Borden of the Fall River Iron Works to construct a mill that was based on the specifications of the Bolton mill. The result was the Metacomet Steam Cotton Mill, built in 1847. Davol and Durfee were the mill's architects.

The Metacomet was the first mill in the city to use steam power. It was originally powered by two breast water wheels 19 feet in diameter and one steam engine of 160 horsepower. Later, the breast water wheels were replaced with two turbine water wheels and its steam power upgraded with a Corliss engine of 375 horsepower, which provided two-thirds of the power of the enlarged mill.

The Metacomet was the first mill in America to use cast iron posts and girders. This innovation resulted in removing the friction caused by the movement of buildings constructed only of wood and therefore resulted in more efficient use of power. About two-thirds of the machinery was American-made, also an innovation during a time when most textile machinery was being imported from England.

The original building was 247 feet long by 70 feet wide and five and one-half stories high with a barn roof. The Metacomet was wider than earlier Fall River mills, another innovation borrowed from the Bolton model. Later, the barn roof was replaced with a flat roof and the half-story replaced with a sixth story. Additions were also made to the building over time.

The rubble granite used in the construction of the Metacomet came from an on-site quarry at the nearby granite cliff. The wall thickness tapers from 3'-0" at the ground floor to 1'-8" on the fifth. The exterior facing is parged, or roughly mortared and smoothed over with a cement mixture. The corner stones and the face of the stair tower and office wing are neatly-shaped, squared ashlar granite. Granite lintels span the openings of doors and windows. (28)

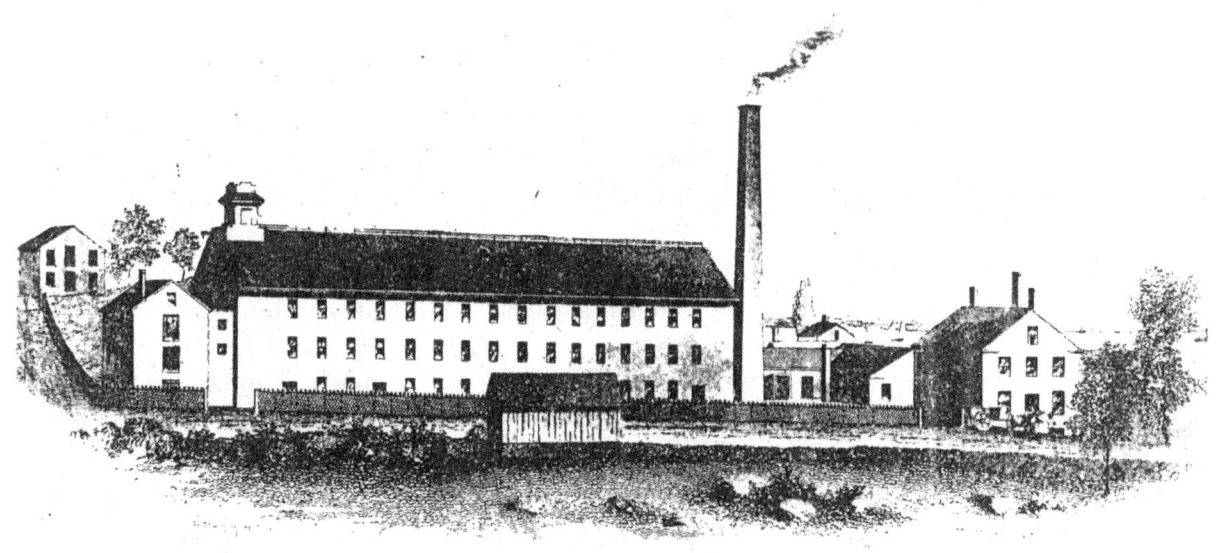

fig. 10.2 **Oliver Chace Thread Mill**

Note the office building to the right, still surviving on Bay Street and shown in the photo below.
Archives of the Fall River Historical Society.

fig. 10.3 **Oliver Chace Thread Mill office building**

This surviving building demonstrates the construction design characteristics of the second phase of textile mill construction that was used in the main Chace Thread Mill building, with its Greek Revival influence, gable roofs, and massive random rubble granite walls with small windows and granite lintels.

new Pocasset Mill, built in 1847, was substantially larger than previous mills, being 219 feet long and five stories high and made of granite taken from the city's newly-opened quarries.

Following a visit to Bolton, England to observe a textile mill reputed to be a model of English engineering, Major Bradford Durfee and William C. Davol collaborated with Jefferson Borden to construct a mill based on the specifications of the Bolton structure. The result was the Metacomet Steam Cotton Mill, built in 1847. The Metacomet was the first steam-powered mill in the city and the first textile mill in America that used cast iron posts and girders. It was equipped as much as possible with American-made machinery, powered by steam and water.

Mills in this phase were typically constructed of random rubble granite with heavily parged joints and corners of squared ashlar granite blocks and granite lintels over doors and windows. They were built with small windows and gable roofs. The Metacomet Mill, located on the Quequechan River falls at the corner of Anawan and Davol Streets, is the only remaining textile mill from this era in Fall River.

The Metacomet took raw cotton and processed it into print cloth. It contained 23,840 spindles and 591 looms and manufactured 6.5 million yards of cloth annually from 2,500 bales of cotton. In 1877, the mill owned 56 tenements to house its operatives. It was part of the Fall River Iron Works complex and was later named American Printing Company Mill No. 6. The brick mill complex east of the Metacomet, known as American Printing Company Mill No. 7, was built in 1905-06. (28)

The Metacomet Mill was recorded in 1968 as part of the New England Textile Mill Survey II, which was sponsored by the Historic American Buildings Survey of the Office of Archaeology and Historic Preservation of the National Park Service and the Smithsonian Institution. (28)

The only other mill building remaining from that period is the Greek Revival office building of the Oliver Chace Thread Mill at 505 Bay Street. It is one of the earliest surviving mill office buildings of that era and shows the wall construction of the Chace Mill that once stood at that site.

The Doctor Nathan Durfee dye manufacturing mill, located on Mill Brook in the Southeastern Massachusetts Bioreserve, also dates from the second era. The exposed ruins of that mill show the granite used in the construction of the walls. The granite for Dr. Durfee's mill was taken from a small quarry in East Fall River.

The third era of mill building

The third major era of mill building in Fall River occurred from 1859 through 1875. These mills are almost always characterized by gable roofs and imposing bell towers. The gable roofs continued the barn characteristics from the earliest mills. Skylights on the gabled roofs provided extra light. The ideal dimension of textile mill buildings during this era of building was 300 feet in length by 72 feet in width.

The first of the third era mills to be built was Union Mill No. 1, located behind Union

fig. 10.4 **Union Mill No. 1**

Union No. 1 can be credited with a number of firsts. It was the first mill to be built above the Quequechan falls, the first mill to be financed by community stock subscriptions, and among the first to be powered entirely by steam. It was also the first to install the Corliss steam engine, where the exhaust steam could be used to dress the loom warps and to heat the building in winter. All of the Union Mills were originally built with hipped roofs, similar to the Durfee Mills in the same yard, but the roofs were later flattened and an additional full story added.

fig. 10.5 **The Durfee Mill complex**

Shown is the office building on Pleasant Street, with the two main mill buildings and towers in the background. This impressive complex of 12 buildings (including the Union Mills) was the subject of a survey conducted by the Smithsonian Institution and the National Park Service in 1968 and 1971. The complex was restored by the Architectural Conservation Trust in 1983, which resulted in the preservation of the important wooden elements integral to the aesthetic character of these remarkable buildings.

Alfred J. Lima

fig. 10.6 **Tecumseh Mill No. 1**

This is perhaps the best example of the beauty of Fall River's textile mills when a full restoration occurs, one that preserves the original wooden window sash, roof overhangs, and brackets. It is also an example of how the narrow mills of the third stage of Fall River's mill development lend themselves to residential reuse. This stage of mill construction still retains the high-pitched roof original to the English barn.

fig. 10.7 **The Crescent Mill**

The mill, at Pleasant Street and Plymouth Avenue, is a fine example of the third mill period in Fall River.

fig. 10.8 **The Chace Mill bell tower**

The granite bell tower of the Chace Mill is one of the most elaborate in Fall River and, along with the Durfee towers, the best preserved. The belfry is characterized by a central arched opening flanked by smaller arches, with caps at the top of the columns and detailed keystone work forming the arches. The cornice includes double brackets and a continuous row of wooden dentils. All of the bell towers in the city appear to be unique, with no two towers alike.

fig. 10.9 **The Chace Mill**

Built in 1872 during the post Civil War building boom, when 22 mills were built in the brief period of 13 months, the Chace Mill is the classic example of Fall River's textile mills of the third period. It is long and narrow (377 feet long by 74 feet wide), high in profile (six stories) and with a prominent stair/bell tower. Along with the Durfee-Union Mills, the Chace was included in the New England Textile Mill Survey of 1968/1971. The Chace Mill is particularly imposing because of its isolated location.

fig. 10.10 **Italian Romanesque campaniles**

Fall River mill towers during the third phase of mill development were modeled after church campaniles that were built during the Romanesque period in Italy during the late Medieval period. The above drawing of Saint Abbondio Church, built in Como, Italy, from 1063 to 1095, shows the distinctive characteristics of these campaniles. The architectural elements that Fall River builders and architects borrowed from these Romanesque towers included their square shape, four-sided hipped roofs, triple arched openings, and bracketing below the roofs and bell openings. Illustration from Fletcher: *A History of Architecture*.

fig. 10.11 **Mill towers in the third phase of mill development**

The Italianate architectural style was very popular in America during the second half of the 19th century, and the bell towers of the Fall River mills built in the third phase reflect that stylistic preference. The mill tower on the left is one of two matching towers at the Durfee Mills at Pleasant Street, with wooden shutters that were unique among Fall River towers. The tower on the right is at the Granite Mills at Bedford and Robeson Streets. Its roof brackets have been removed. Bell towers called operatives to work and contained the stairways and hoists that provided access between floors for workers and material.

Mill No. 2 on Pleasant Street. Union No. 1 is a transitional structure from the second to the third era of mill building. It was built in 1859 and is about half the size of what would become the typical phase-three mill. It measures 160 feet x 70 feet and had a capacity of 15,000 spindles. Its construction was of irregular granite ashlar with parged joints used in the mills of the earlier era. (28)

Union No. 1 can be credited with a number of firsts. It was the first mill to be built above the Quequechan falls, the first mill to be financed by community stock subscriptions, and among the first to be powered entirely by steam. It was also the first to install the Corliss steam engine, where the exhaust steam could be used to dress the loom warps and to heat the building in winter. (28) This mill has recently been handsomely restored and is the home of Prima Care.

All three of the Union Mills were originally built with the gabled roofs of the third period, but all three roofs were later lowered to the flat roofs of the fourth period and the top half-floor converted to a full sixth floor. The Richard Borden Mill, at the corner of Rodman Street and Plymouth Avenue, was the first mill in the third era to be built with a flat roof, a transition into the fourth era.

The unique bell towers of the third period mill buildings have distinctive design and detailing. One of the most elaborate towers is part of the Chace Mill on Rodman Street. The belfry has a central arched opening flanked by smaller arches and terminates with a row of wooden dentils, a double-bracketed cornice, and a hipped roof. Virtually all of the tower archways were open, with the exception of the two Durfee Mill towers, which were built with wooden shutters in the tower openings. This third major period of mill construction occurred during the time when the Italianate architectural style was prominent, and the towers in particular reflect the Italianate influence. These towers have a striking resemblance to campaniles in Italian Medieval hill towns, for example in Assisi.

The third era of mill building is best exemplified by the Durfee-Union Mills at Pleasant Street and Plymouth Avenue. The central grouping of the Durfee Mills is formed by two 5-1/2 story Italianate mills with handsome 7-level center towers, erected in 1866 and 1871.

The buildings are placed symmetrically at right angles to Pleasant Street, forming a large center court that was originally landscaped. In the front center of the courtyard is the 2-1/2-story granite office, built in 1872. (28)

The Durfee-Union complex includes about 12 buildings, whose main structures were begun in 1866 and completed in 1872, surrounded by an iron fence. In the earlier periods of mill construction, granite blocks were hoisted first on ramps and later hoisted by power provided by oxen. In 1866, the Durfee Mills were the first to use portable steam engines to hoist granite in Fall River, a method that was found to be "much more economical than horseflesh." (52-188)

The Durfee-Union complex was restored in 1983 by the Architectural Conservation Trust, a private company with a focus on rehabilitating architecturally significant properties.

The Durfee-Union Mills complex was recorded as part of the New England Textile Mill

fig. 10.12 **The Arkwright Mill and Davis Mill No. 2**

The Arkwright (left) and Davis No. 2 with the Quequechan River floodplan/wetlands in the foreground. The high-pitched English barn roof of the third phase has given way, in the fourth phase, to a wide, low pitched roof.

fig. 10.13 **Davol Mill No. 2: A classic example of mills in the fourth phase**

The most singular characteristic of the mills in the fourth phase of textile mill development in Fall River is the high ratio of glass in relation to granite, giving the mill buildings in this phase a remarkable lightness of construction and an unusual aesthetic. In addition to having more glass surface, the mills in the fourth phase are much wider than the third phase and are lower, having fewer floors. Mills in this phase typically have two or three floors. Other characteristics of fourth phase mills include low-pitched roofs, lack of ornamentation found in the third phase, and wide overhanging cornices with single wooden brackets. Granite blocks are generally rough cut and set in random courses with parged joints. More finished blocks are used only for corner quoins and lintels. A new fenestration also emerged in the fourth phase, with heavy cross mullions dividing the windows into four parts. This Davis façade faces Quequechan Street.

Survey II, which was sponsored by the Historic American Buildings Survey of the Office of Archeology and Historic Preservation of the National Park Service and the Smithsonian Institution. The project was assisted by the Merrimack Valley Textile Museum in North Andover, Massachusetts. Field work, historical research and record drawings were conducted in 1968, with historic documentation and editing of the project data conducted in 1971 under the auspices of the Historic American Engineering Record (HAER) of the Office of Archeology and Historic Preservation of the National Park Service. (28)

The third phase of mill development is also significant because of the rapidity with which the mills were built. During a period of only 13 months—from February 1871 to April 1872—a total of 22 mills were built in the city, an extraordinary accomplishment. (73-51)

Fourth era of mill building

The fourth era of mill building, concentrated in the period from 1880 to 1910, is quite unique because the structures built in this era were designed and constructed with a very high ratio of windows and a relatively small amount of granite support structure. Because of the high ratio of glass, they give the impression of a remarkable lightness of construction.

Textile mills in Fall River built during this period show a marked increase in width, which allowed for better placement of machinery and more efficient operation of industrial processes. Unlike the mills of the previous phase of the 1860s and 1870s, where the width of mills in the city was rarely more than 72 feet, the mills in the fourth phase of mill construction were much wider. For example, the Hargraves mills (1892 and 1893) are 127 feet wide; the Parker (1895) is 150 feet wide; the Arkwright (1897) is 125 feet wide; and the two Davis mills (1902 and 1908) are a full 160 feet wide. (28)

The wider mill design, however, required that windows be larger to let in more light into the interior of the buildings. In these mills, there is far more window than granite. The expanse of window and the delicacy of the sash—sometimes with as many as 60 panes of glass in each window—give a striking effect of lightness. This is truly remarkable, given the amount of vibrating machinery that operated in each building.

As a general rule, for both granite and brick mills in Fall River, the later the mill, the higher the ratio of glass to stone or brick.

Characteristic design features of the fourth mill phase include low-pitched roofs with wide overhanging cornices with single wooden brackets. Granite blocks were generally rough cut and set in random courses with parged joints. More finished blocks were used only for corner quoins and lintels. The large windows are characterized by 6/18 and 6/24 T-cross sashes. (28) Unlike the earlier mill construction phase, the fourth period of mill construction included no towers.

Also absent were gable roofs, which were flattened. The Granite Mill fire of 1874 demonstrated the hazard of that construction. In that fire, employees on the top or gabled floor could not access fire escapes, which were on the long sides of the building.

Two of the city's mills of this era have buttresses. The Connell Mill on Alden Street

fig. 10.14 **The Durfee Mills weaving shed**

Added to the Durfee Mills complex later, this weaving shed is one of the most well-preserved examples of the fourth phase of mill construction in Fall River. It demonstrates the importance of preserving the original fenestration of the fourth phase mills, with its wooden cross mullions and delicate sash creating a well-proportioned contrast with the granite mills.

fig. 10.15 **The Seaconnet Mills**

This is another example of the lightness of fourth phase mill construction on the Quequechan River, where the building seems to consist of all window and barely any granite wall. This mill was recently renovated, with the original fenestration (formerly removed and blocked in) replaced with windows that reproduce the effect of the original.

fig. 10.16 Characteristic mill chimney

One of the more beautiful characteristics of Fall River's mill buildings were the fluted brick chimneys that were once visible everywhere in the city. These chimneys were characterized by a broad base tapering upward to the top, then flaring out and in again at the top opening. The aesthetics and workmanship of these chimneys is extraordinary. Only one of these chimneys remains intact at the Darwood Mill on Pocasset Street, although the visual effect has been marred by cell equipment. Note the gothic window openings of the American Printing Company Mill No. 7, a distinction not repeated in any other mill in Fall River.

fig. 10.17 Mill workmanship

One of the two impressive bell towers at the Durfee-Union Mill complex, beautifully designed and built, still perfectly intact after 140 years. Historian Sylvia Chace Lintner notes that much of the credit for the excellent construction of these mill buildings must go to the supervising masons and carpenters. "Much that is good about the Fall River mills," she says, "must be credited to these men, for they were craftsmen of a high order, with a fine regard for honest workmanship and a respect for their trade much like that of George Eliot's Adam Beade. The solidity of construction and the skilled handling of great blocks of granite draw admiration today." (35-192)

has six buttresses in regularly-repeating bays on each principal facade in a 3/6/6/6/6/3 pattern. The Hargraves Mill on Quarry Street also has buttresses. Andrew Roy, who was raised in the Flint, says that the reason for the buttresses was the movement of the shuttles on the looms, going back and forth, could cause the mill buildings to crack. (90)

The Arkwright and Davis mills on Quequechan Street are among the best examples of this last phase of mill building. All are the work of Chauncy Sears, a mason and builder who specialized in mills.

These large mills of the later period (Davis Mills No. 1 and No. 2 measure a full 355 feet x 160 feet and 435 feet x 160 feet) were distinguished by their large spindle capacity and specialized product line and also because they remained in manufacturing use longer than most of Fall River's mills.

Millyards often include more than one example of textile mill architecture, because mill corporations often added to their plants as the need arose and as economic cycles dictated. The Durfee-Union Millyard, for example, includes building styles from the second, third,

fig. 10.18 **Mechanics Mill**

The Mechanics Mill on Davol Street represents perhaps the finest example of the brick textile mills that were built below the hill on the Taunton River and Mount Hope Bay. It was designated in the Second Empire architectural style that was popular following the Civil War.

Reproduces from Earl: *A Centennial History of Fall River, Massachusetts*

and fourth mill periods, with the second predominating.

The fourth phase includes many free-standing weaving sheds that were added later to mill complexes, including the Durfee-Union Millyard and other mills. Where these weaving sheds added to increase capacity or were they built to remove the stress on the main buildings from the vibrating looms?

The Geographic Distribution of Granite and Brick Mills in Fall River

Fall River's mill buildings are built of either granite or brick, and they show a distinctive geographic distribution. Granite textile mills predominate above the hill and the Quequechan River falls, on the flatlands surrounding the flooded upper river. Below the hill, however, on the shores of the Taunton River, brick mills predominate.

This interesting geographic distribution of mill buildings—granite above the hill and brick below the hill—exists because it was less expensive to transport bricks via freighters from kilns in Providence or Taunton, two major local sources of brick. In addition, before 1875, when the Watuppa branch of the Old Colony Railroad was built to serve the industrial area above the Quequechan River dam, there was no cost-efficient way to bring brick to the area above the falls.

The city's several granite quarries, however, were located above the hill, sometimes on the mill property itself. Perhaps another reason for the lack of granite below the hill was the difficulty of transporting large granite blocks down the steep hillside before the advent of internal combustion engines. Virtually all of the mills in the vast complex of the Fall River Iron Works and the American Print Works, at the foot of the Quequechan River falls, were built of brick. On the upper reaches of the falls, the Troy and the Pocasset mills—now gone—were built of granite.

One of the curious exceptions to this distribution of brick and granite is the Sagamore Mill complex, located between North Main Street and the Taunton River, two of whose mills are built of granite. Sagamore No. 2 is built on North Main Street at Cove Street. The answer to this puzzle is that the granite was brought in by rail via a spur track from a new quarry in Assonet opened especially for supplying granite to the two Sagamore mills. (28) This is the only one of the area's former granite quarries that is clearly visible today and is now preserved as part of Freetown-Fall River State Forest and the Southeastern Massachusetts Bioreserve.

Another exception to the distribution pattern is the Metacomet Mill, built of granite in 1847 at the base of the Quequechan River falls. In this instance, the Metacomet was built of granite quarried from the ledge on the hill abutting the mill.

Good examples of brick mills that can be seen today are the Mechanics Mill on Davol Street, built in 1867. The etching on the previous page shows the original design and construction of the mill, with its unusual octagonal bell tower and Second Empire roof design. The Narragansett Mill and the Border City Mills on North Main Street are also excellent examples of the brick mill style common during the third phase of mill construction in the city. These were also designed in the Second Empire architectural style that was popular at the time.

How Architecture Reflected the Division of Work in the Mills

In his 1877 *Centennial History of Fall River, Massachusetts*, Henry H. Earl says that the ideal printing mill in the city had certain proportions:

> The experience of the Fall River cotton manufacturers has led them to the conclusion that the most desirable size of a mill, for the manufacture of print cloths, is one of 30,000 spindles. In such a mill, the different parts balance each other to the best advantage; that is, if properly arranged, the looms will just take care of the preparation—the carding, spinning, dressing, etc.—with no surplus or deficiency. It is also about as large as a superintendent can handle easily, by keeping up the different ends, and having every thing run smoothly, without hitch or break. (25-110)

Earl spoke of a mill of phase three design, the dimensions of which were generally 300 feet long by 72 feet wide. It was five stories high with a hip or flat roof, "the latter more desirable because of fire." Its capacity was 30,000 spindles and 800 looms and would employ from 325 to 350 operatives. It would use, per annum, 3,500 bales of cotton and produce 9,000,000 yards of print cloth. Four to 10 acres were required for a mill site, depending on how much housing was provided for workers. In 1877, capital of about $500,000 would be needed to erect the building and equip it with machinery, with a small margin for working capital. The building would account for about 40 percent and the machinery 60 percent of the total cost. (25-110)

The arrangement of the machinery in the building was important to the proper functioning of the whole operation. The first and second floors were used for weaving, the third for carding, and the fourth and fifth for spinning. The steam engine room was in an ell in the center rear of the building. The main driving wheel and shaft were located in the basement, from which the belts transmitted power to the various floors. The front central tower provided access and accommodated the bell that rang the operatives to work. (25-110)

In her 1948 article in the New England Quarterly titled "Mill Architecture in Fall River," Sylvia Chace Lintner notes:

> The interior design of a mill depended upon the manufacturing processes. The typical Fall River factory had the heavy looms installed in the first two floors. The third floor was used for carding because it was on a level with the picker room, which was isolated in a rear ell to prevent the spread of fire. A forced draft blew the loose cotton through a tube from ell to carding room. From there it went to the fourth floor for spinning and to the attic for spooling, dressing and

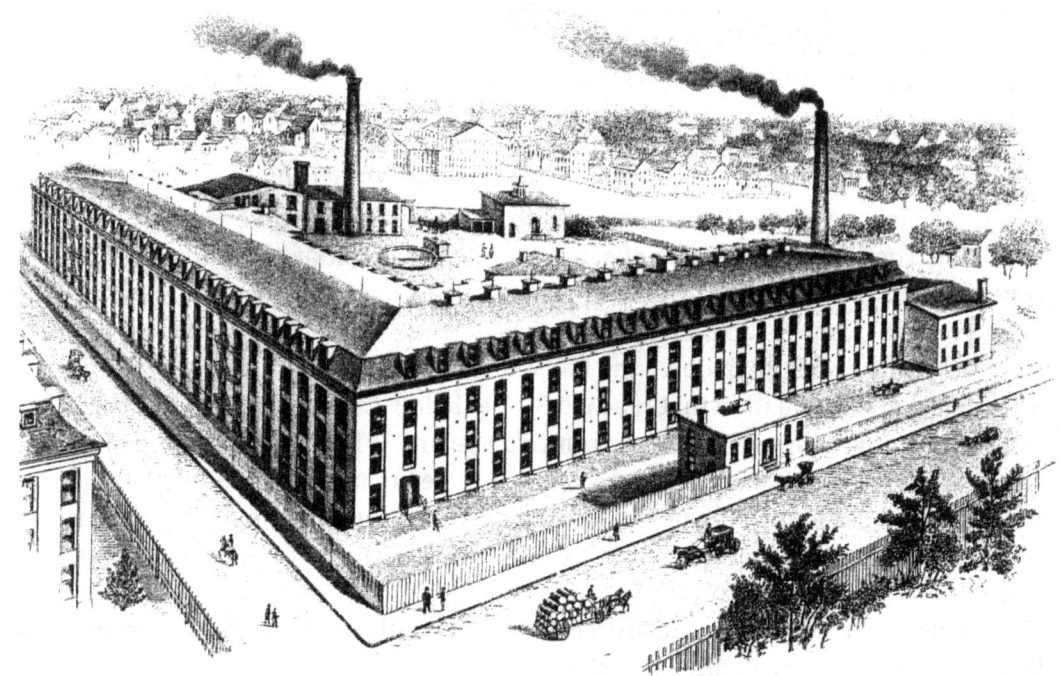

fig. 10.19 **Davol Mills**

At the corner of Rodman and Hartwell Streets. One of the few brick mills built above the Quequechan River falls, the Davol was designed in the Second Empire architectural style, which was popular in the era following the Civil War. The mill is unusual for its full Mansard roof, typical for the Second Empire style but rare for Fall River mills. Most of the city's early brick mills were built in the Second Empire style.

Illustration from Earl: *A Centennial History of Fall River*

warping. Dressing, a process which involves sizing the warps in steaming vats, was done on the top floor where the heat could not permeate to the rest of the building. Separate structures or ells attached to the rear of the main building provided the extra space necessary to house the engine, the boiler, the picker room, a cloth room, and cotton storage. (52-193)

The vibration caused by heavy machinery was no small matter for textile manufacturers. When the Pemberton Mill collapsed and burned in Lawrence, the issue came to public attention and raised concerns in the industry. The oscillation in the upper story of one Fall River mill was four inches or more and once so alarmed the help that they left the building fearing for their lives. If filled to within six or eight inches of the top, barrels of water in the top floor would spill over. (25-32) Benjamin Pearce, who worked in the Fall River mills as a boy, tells of a similar experience in his memoirs:

The looms were in the third story and swung laterally of the mill. At times during the day, these would assimilate the lock-step of marching, and swinging in unison would cause the huge structure to sway perceptibly at the top. At one time the help left the mill in a panic, fearing that it was about to fall. (62-37)

During his visit to Fall River from Manchester, England, in 1902, T.M. Young observed "the way in which the looms dance and jump upon the wooden floors." (88-11)

The construction of the mills reflected the concerns about the effects on building integrity posed by stress from constant vibrations. This began with firm foundations. When building over the Quequechan falls, care had to be taken that the vibration would not result in buildings gradually slipping down the hill. The new Pocasset Mill, built directly over the river in 1878, had a foundation that reached 50 feet below the first floor on one corner and bridged the river by heavy timbers on strong iron beams. When building along the edge of the upper muddy Quequechan River, care had to be taken that the foundations were on firm ground. Lintner mentions that when the Union No. 3 was built on the banks of the river in 1877, over 7,000 tons of granite cobble were dumped in the river to assure an adequate base for the foundation. (52-193)

The walls of the mills reflected the stress inside, with walls thicker at the first floor and tapering to the upper floors. Lintner notes that the usual type of mill construction is illustrated by Union Mill No. 1:

> The walls on the first floor were three feet, three inches thick; those of the second floor, three feet; the third floor, two feet, eight inches; the fourth floor, two feet four inches; and the fifth two feet—all 'laid with cement and lime mortar in the best possible manner.' The work required 1,000 barrels of lime and 700 barrels of cement. The floor timbers were twelve by sixteen inches, sometimes seventy feet long, and supported by heavy iron columns based on thick subterranean piers. Between each story was a layer of solid construction six and a quarter inches thick, composed of four-inch plank with covering and inch and a quarter and a one-inch ceiling below. (52-194)

The Creators of the Mills: Architects, Engineers, Builders, and Their Clients

The early days of mill construction were the product of a few men of independence and remarkable initiative. As Sylvia Chace Lintner notes:

> A clear line of influence extends from the architects of the heyday of the city back to the experience of the earliest builders. The developments characteristic of Fall River mill design can hardly be understood without reviewing the work of the few men chiefly responsible for it. Those who reared the first structures when the city was only a small farming community were among the original entrepreneurs, men of independence and initiative. All of them had experience

as mechanics, and their approach was that of the practical builder. Their activity has a familiar pattern: a little contact with others in the same area, independent effort, importations from England, further independent experience, and finally professional training for the generation which inherited their work. (52-186)

The Fall River textile industry and its mills began with a pattern of experimentation with modest manufacturing enterprises on small streams in surrounding towns as part of a farm operation. This is particularly true of towns between Pawtucket and Fall River, including Rehoboth, Swansea, and Somerset.

David Anthony of Somerset had apprenticed with Slater and the Wilkinson brothers in Pawtucket and later joined in a mill enterprise in Rehoboth initiated by his cousin Dexter Wheeler. Wheeler had a blacksmith shop, grist mill, and mill privilege on his farm in Rehoboth. Anthony and Wheeler joined together to form the Fall River Manufactory. Wheeler was an experienced mechanic who, in 1813, built all of the machinery for the mill, including the first power looms used in Fall River, introduced in 1817. (25-12,13,16 and 119)

fig. 10.20 **The Merchants Mill**

One of the more impressive and imposing of Fall River's textile mills, the Merchants Mill occupied the entire block between 13th, 14th, Pleasant, and Bedford Street. The Merchants was designed by Lazarus Borden in 1866 and was built of granite quarried on the site at the Bigberry Ledge, with additional granite supplied from Beattie's quarry. It was one of the few granite mills in Fall River built in the Second Empire architectural style; most of the granite mills of the third period were designed in the Italianate style.

Illustration reproduced from Earl: *A Centennial History of Fall River, Massachusetts*

Alfred J. Lima

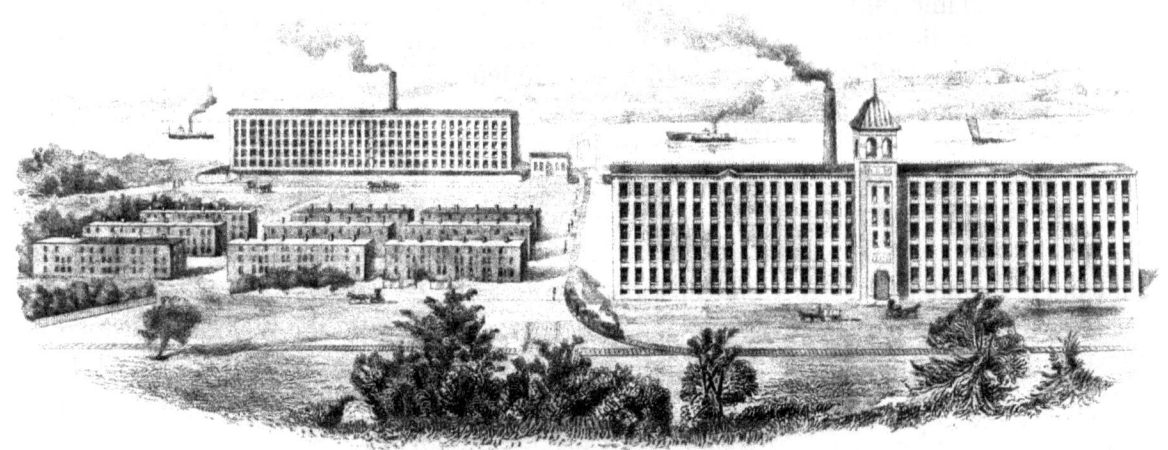

fig. 10.21 **The Border City Mills**

The Border City mills were designed in the Second Empire architectural style by Josiah Brown, the first of the professionally-trained mill designers. Mentored by William C. Davol, Brown was mostly self-taught and traveled to Europe for study and observation. The Border City structures are among the brick mills that were built along the Taunton River from bricks brought by ship from either Taunton or Providence. The workers' housing shown on the left were later demolished for new development.

Illustration from Earl: *A Centennial History of Fall River*

Oliver Chace of Swansea was a carpenter and wheelwright who acquired a knowledge of spinning by working as a mill-hand and had joined with Dexter Wheeler in building and operating a small factory in Rehoboth in 1806. Chace later formed the Troy Cotton and Woolen Manufactury in 1814, the second textile manufacturing enterprise on the Quequechan River.

Major Bradford Durfee was the most active of the early builders, who began his career as a blacksmith and shipbuilder. With Colonel Richard Borden, he established the Iron Works in 1821 and built its first buildings at the base of the Quequechan falls. As long as fieldstone was used for building the mills, stone could be brought to the masons on wheelbarrows or rolled up ramps. The introduction of granite required more horsepower, and Durfee excited local admiration by introducing the use of oxen to deliver and hoist granite to the Anawan Mill (1825), the first mill in the city to be built of granite (although the top two floors were built of brick). In 1839, Major Durfee also built a new mill for the Fall River Manufactory, replacing Anthony's original building. (52-187)

Major Durfee's trip to England in 1838 with his nephew, William C. Davol, resulted in innovations that would begin the second phase of mill development in Fall River. Upon their return, Durfee and Davol designed the Metacomet Mill, based on the famous cotton mill in Bolton, England. This 1843 mill was twice as wide as earlier mills, allowing for the better placement of spinning machines, and was built with iron posts and girders, an innovation in America. In the Metacomet, Durfee and Davol introduced steam power to Fall River textile manufacturing. (25-57)

Stephen Davol built on the accomplishments of Bradford Durfee and Stephens's older brother, William Davol. Stephen started as a doffer boy at the Troy Mill and rose to superintendent of the Pocasset in 1833. He brought a systematic method to planning for mill design, using sectional drawings to get the best belt connections. His work elicited much admiration by his peers in the industry. (52-189)

Following 1859, the city's mills were created by men who were professionally trained as architects or engineers. The first of the professional designers was Josiah Brown, who was mostly self-taught, traveling to Europe for study and observation. William C. Davol mentored Brown in his early mill work, principally on Union Mill No. 1, built in 1859. He soon built himself a reputation and developed an office of assistants trained in his methods. Among his accomplishments were the Border City Mills. (52-189)

One of Josiah Brown's cadre of assistants was William T. Henry, perhaps the greatest of local mill architects. Henry graduated from MIT in 1870 and eventually became head of Brown's office in 1876, several years after Brown left to become chief construction engineer for the Hoosac Tunnel. William T. Henry built 46 Fall River cotton mills between the late 1870s and the early 1900s.

David Hartwell Dyer, another of the mill designers, had been closely involved with the local textile industry all of his life, having been an officer of several of the city's mills in various capacities. Among the mills that he designed were the Weetamoe, Sagamore, Flint, and the Osborn.

The role between designer and client was often blurred. As Lintner observed:

> Often the rough plans and drawings for a new mill were prepared by the treasurer, who then consulted a builder. The chief concern in constructing a cotton mill is utilitarian, and the men in charge needed to be sure that the design was well adapted to cotton processing. Thus, in the case of the Granite No. 1, built in 1863-1864, Charles O. Shove, who was both agent and treasurer, supervised the planning, erection, and arrangement of the building, with the help of Lazarus Borden, who was often engaged to 'furnish the drafts, contract for the machinery and see to the general "winding-up" of other Fall River mills.' (52-191)

One of the more impressive and imposing textile mills in Fall River was the Merchants Mill, designed by Lazarus Borden in 1866. However, little is known of the training that Borden may or may not have received in architecture or engineering.

Thomas J. Borden, treasurer and corporation clerk of the Richard Borden Manufacturing Company, planned and "wound up" the Richard Borden mill himself in 1872. Thomas had two years of study at the Lawrence Scientific School in Cambridge, and the mill he designed included innovations such as double towers and a nearly flat roof. The Richard Borden Mill was considered to be "one of the most perfect structures for manufacturing purposes in the country." (25-47) The creator of the Mechanics Mill on Davol Street is not known, but Lintner says that any one of the organizers of the mill—Stephen Davol, Lazarus Borden, Thomas J. Borden, and D.H. Dyer—would have been capable of designing the mill.

Finally, much credit is due to the supervising masons and carpenters responsible for the actual construction of the mills:

> Much that is good about the Fall River mills must be credited to these men, for they were craftsmen of a high order with a fine regard for honest workmanship and a respect for their trade much like that of George Eliot's Adam Bede. The solidity of construction and the skilled handling of great blocks of granite draw admiration today. (52-192)

Preserving the Architectural Integrity of the Remaining Mills

Preserving the architectural integrity of Fall River's textile mills depends in great part on finding appropriate economic uses for them. The wide and long mills of the fourth period of mill design have generally remained in textile manufacturing use until recent times. Their large floor plans allow the flexibility of machinery layout that the narrower mills, built in earlier periods, do not.

The narrower mills of the third period lend themselves to conversion to non-manufacturing uses, including retailing, office, and residential use. The conversion of the Tecumseh Mill on Hartwell Street to residential use is an excellent example of mill restoration, as is the Border City Mill in the North End and the Wampanoag Mill on Quequechan Street. The Durfee-Union Millyard represents an ambitious restoration project that succeeds because of its mixture of restaurant, retail, small manufacturing, office, and other uses. The recent restoration of Union Mill No. 1 into a medical office building is a perfect fit for a mill of this size. The Narragansett Mill on North Main Street has been handsomely restored for office use. The brick Mechanics Mill, has recently been converted to new uses that include offices and restaurants.

What makes Fall River architecture unique is its granite mills. These are truly extraordinary structures. They have literally emerged out of the ground, out of the granite foundation that underlies the entire city. Perhaps no other city can claim that distinction.

Preserving the integrity of the granite mills requires that the vulnerable wooden sections of the mills be saved and restored: specifically, window fenestration, eves, cornices and brackets. The original windows give the mills the contrast between the solidity of the granite contrasted with the delicacy of small panes of the original fenestration. The overhang of the eves and their brackets provides a shadow line that defines the roof line and makes it stand out against the sky.

The projecting eve of Union Mill No. 1 creates a shadow line, making the building stand out against the sky. The decorative brackets support the extended eves and also emphasize the vertical rows of windows. The original wooden window frames and sash complete the aesthetic of the whole composition.

One of the Durfee Mill buildings, showing the shadow lines created by the overhanging eves on the building and the bell tower. This example illustrates the importance of preserving original window casings and decorative brackets at the eves. Many bell towers in the city have been altered to remove their eves, resulting in a truncated effect. The eves are important not only for the shadow line that they provide but also because they compete the image of the Italian late Medieval campaniles on which they are modeled. The belfry shutters and the original doors on each floor have also been preserved on the Durfee towers.

fig. 10.22 **Elements integral to the preservation of the architectural integrity of Fall River's mill buildings**

Too many of the city's granite and brick mill buildings have been demolished or destroyed by fire. The recent reuse of mill buildings shows that they can have a continuing and valuable economic use. However, these new economic uses succeed best when they respect the original architectural integrity of these unique buildings.

11 The Fall River Iron Works and its Subsidiary Enterprises

The Fall River Iron Works

While an iron manufacturing enterprise may seem oddly out of place in a town so singularly focused on the manufacture of textiles, the Fall River Iron Works was to prove invaluable in the development of the city's textile industry. Without the Iron Works, a smaller amount of capital would have been available for textile mill expansion. In addition, it diversified the local economy and stimulated the development of a significant local textile machinery industry, which proved to be an important advantage for cotton textile manufacturing in the city. (73-38)

By its advancement of steamship connections to New York City and by the construction of railroads to the city, the Iron Works assured that the textile industry was served by a first-rate transportation infrastructure. Not least, the Iron Works own investments in textile finishing and printing (through the American Print Works) complemented the main business of other textile manufacturers in Fall River, whose focus was in converting cotton into gray cloth.

One of the more neglected chapters of the history of Southeastern Massachusetts is its location as an early iron center. Taunton (now Raynham) was the site of one of the earliest iron works in New England, established in 1656.

Iron works began operating along Fall Brook in East Freetown from early Colonial times, when iron ore was taken from the bogs at Assawompset Pond, with fuel available from East Freetown forests. When the foundry was enlarged and updated in 1818, higher quality ore and pig iron from New Jersey was brought up the Taunton and Assonet Rivers by sailing vessels, then unloaded on Assonet wharves. Teams of oxen would then bring the iron ore and fuel to the furnaces in East Freetown, to a settlement soon called "Furnace Village." Later, foundries were built on the Assonet River at Forge Pond and on Terry and Rattlesnake Brooks. (4)

In the *History of the Town of Dighton, Massachusetts* by Helen H. Lane, a reference is made to the forge at Westville on the Three Mile River, a major tributary of the Taunton River at the Taunton/Dighton corporate line, operated by Hodijah Baylies. Baylies' forge fabricated the huge anchor for the USS Constitution. Ten yoke of oxen were required to bring the anchor overland to South Dighton, then the highest point of navigation for major ships, where it was loaded on a ship and brought down the Taunton River to Boston. (50)

The Fall River Iron Works had its genesis in shipbuilding and with two men who were to have a major impact on the history of Fall River: Major Bradford Durfee and Colonel Richard Borden. Bradford Durfee had been a ship's carpenter in New Bedford. When he returned to Fall River, he and Richard Borden operated a grist mill together at the foot of the Quequechan River and also constructed small vessels there. In addition, they experimented at the nearby blacksmith shop and gradually established a business manufacturing spikes, bars, and rods. The business grew and, in April, 1821, the Fall River Iron Works was established by Major Durfee, Colonel Borden, and six other stockholders holding equal shares. (25-48)

Major Durfee was in charge of the manufacturing and Colonel Borden did the selling. At first, the major product of the Iron Works was iron hoops for New Bedford whale oil casks. However, as the quality of the firm's nails became known, Richard Borden would assemble a cargo and sail to New York and up the Hudson River until the nails were sold. (25-45)

The first building of the Iron Works was located on the lowest falls of the Quequechan River (where the Metacomet Mill is now located) in order to use the river's power to operate the bellows for the furnace and forge, to raise the trip hammer, and to power the rollers of the slitting mill.

In 1821, a furnace, forge, nail mill, and slitting mill were erected at the Quequechan River falls site. In 1835, another nail factory was built. More buildings were added at this location for the American Print Works.

By 1830, it became clear that the Iron Works site at the falls was inadequate. It was too small for the rapidly growing business and forces in the iron industry were making it necessary to move. About 1815, local bog iron ore and charcoal were becoming increasingly difficult to obtain and therefore more expensive due to the depletion of both iron ore

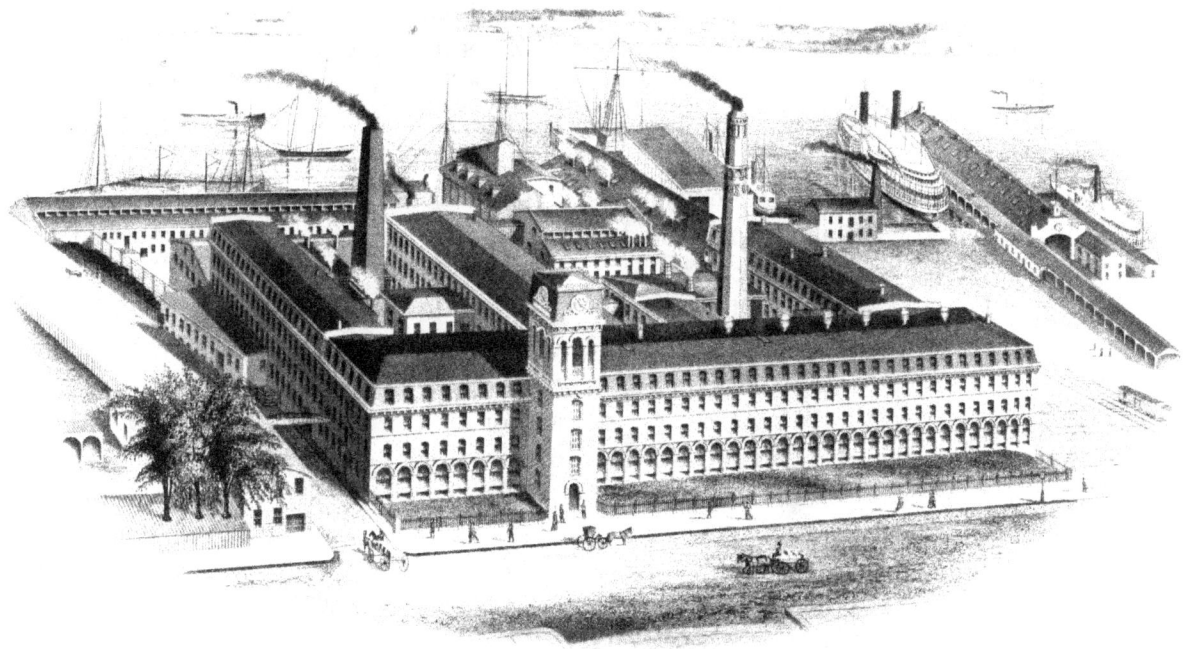

fig. 11.1 **American Print Works and the Fall River Iron Works, 1877**

The American Print Works grew out of the Fall River Iron Works, both shown at this site at the mouth of the Quequechan River where the Taunton River meets Mount Hope Bay. To the right is the steamship pier, where the Providence and New York passenger and freight steamers berthed. The Iron Works Group initiated both steamship companies.

Illustration from Earl: *A Centennial History of Fall River*

and the forests that supplied the charcoal. As a result of this scarcity, New England iron manufacturers turned to abundant and higher-quality "rock" or "mountain" ore from New Jersey. In addition, traditional blast furnaces were being replaced by newer generations of furnaces that used coke or coal instead of charcoal for fuel and that used scrap iron as part of the iron manufacturing process.

These new developments in iron manufacturing now made it essential to have not only large amounts of water power but also large sites, economic handling of bulky materials, and access to water transportation. While the location on the Quequechan River provided water power, the rapid growth of the Iron Works now required a larger site with wharves on Mount Hope Bay.

But how could this be accomplished? The falls were a fixed feature and, while in proximity to tidewater, were still too far from the wharves to allow efficient movement of growing imports of ore and pig iron and exports of finished goods. The resourceful management of the Iron Works resolved this problem imaginatively—by moving the falls. They purchased all of the land extending from the falls to the bay and relocated the Quequechan River to the site. The company did this by damming a saltwater inlet at the bottom of the falls and creating a fresh water pond, what is now known as Crab Pond. It then diverted part of the river to Crab Pond—via a canal constructed by Major Bradford Durfee—and created a fresh water reservoir. A new dam and falls was then built on the relocated site of the Iron Works. (25-46) By this creative and ingenious method, the Iron Works was able to move the lower Quequechan River falls almost one-half mile to the larger iron manufacturing site and its shipping wharves on Mount Hope Bay.

In 1844, the Iron Works rolling mill and nail factory were moved to the new tidewater location. The buildings on the old site on the Quequechan falls were torn down and replaced by the Metacomet Steam Cotton Mill in 1847. (25-44)

Fall River proved to be an excellent location for an iron foundry. After 1815, inland smelters and foundries found themselves at an economic disadvantage. They now had to rely on imported ore and fuel, which had to be brought inland by teams of oxen. Fall River's location—where water power was located at tidewater—allowed the convenient movement of raw and finished iron products by ship and gave the Iron Works site a distinct advantage for iron manufacturing. (73-27)

Colonel Richard Borden

fig. 11.2 **Colonel Richard Borden**

With Major Bradford Durfee, Richard Borden established the Fall River Iron Works in 1821. Richard and Jefferson Borden built the Iron Works into a major industrial empire that laid the groundwork for Fall River's industrialization. Richard Borden established the family's domination of Fall River's textile industry in the 1820s, and with the profits from these textile ventures invested in steamships to Providence and New York and railroads to Boston. Richard Borden founded five of the city's largest mills and was a director of two more. He was a director of two banks, the Iron Works, the Watuppa Reservoir Company, the furnace company, and the gas company. In addition, he held extensive interests in the Providence and New York steamboat companies, railroads, and Maryland coal mines. Richard Borden was instrumental in redrawing the state lines between Massachusetts and Rhode Island that resulted in Fall River substantially increasing its territorial area. As a member of the Electoral College in 1864, Colonel Borden cast a ballot for Abraham Lincoln. Richard and his brother Jefferson grew up on their family's farm in Fall River, which was located between the Quequechan River and Rodman Street, where the Chace Mill and Richard Borden Mills are now located.

Jefferson Borden

Jefferson Borden, born in 1801, and younger brother of Richard Borden, presided over the growth of the Fall River Iron Works and the American Print Works during their periods of greatest expansion. When the Print Works needed more cloth to feed its printing operations, Jefferson built new textile mills. Through judicious borrowing, he managed to save the Iron Works during the financial panic of 1857. He experimented with steam during its early years in the Metacomet Steam Mill. Jefferson managed the American Print Works for 39 years while serving as president of five textile corporations, the Iron Works, the Borden mining company, and a bank. He organized the Metacomet Bank in 1854, then the largest bank in Massachusetts outside of Boston, with capital from Iron Works stockholders. He also initiated the practice of having Print Works agents in New York selling houses. Jefferson Borden was responsible for building a market for cheap printed cloths to replace costly dress goods and by reaching all classes of consumers by a constant change in designs and styles. These seasonally-changing styles became known as "American Prints."

fig. 11.3 **Jefferson Borden**

The Iron Works took full advantage of its coastal location. It imported raw materials such as scrap iron from other coastal cities and bar iron from Sweden and Russia in the early years and later from Philadelphia and Birmingham, Alabama. As the Iron Works converted to steam in the 1840s, its need for coal increased. Coal was brought by water, principally from Nova Scotia, until 1849, when the company bought land at Frostburg, Maryland, and opened a coal mine there.

In addition to making hoops for New Bedford's whale oil casks, the Iron Works manufactured rolled iron for the region's shipbuilding industry and for general purposes, castings for machinery, and nails, and spikes. As early as 1840, Orin Fowler in his *History of Fall River* states that the Iron Works, at its original site on the Quequechan River falls, included two rolling and slitting mills, 42 machines for cutting nails of all sizes, and a foundry for creating castings. In that year, the Iron Works manufactured 38,441 casks of nails of 100 pounds each, or 3,844,100 pounds, 950 tons of barrel hoops and round and square iron, 250 tons of shapes and rods from bar iron, and 400 tons of castings. (38-35)

In 1846, the manufacture and repair of steam boilers was added to the list of Iron Works' industries. In that same year, the corporation started the gas works and began a distribution system for the city. In the erection of gas works, Fall River was two or three years ahead of Providence and several years ahead of New Bedford. (29a-558) In 1856, a department for the manufacture and repair of machinery was added to the Iron Works. During the Civil War, the company held government contracts for the manufacture of arms and other metal products needed for the war effort.

While the Iron Works most important markets were within 20 miles of Fall River, its market area also included New York City and up the Hudson River Valley, with its access to the expanding west. In 1829, two years after the Iron Works initiated regular steamboat service to Providence from Fall River, Jefferson Borden, brother to Richard, was appointed agent to establish an office and warehouse in that city. (29a-557) As its Providence business grew, the Iron Works built a large brick building on the Providence waterfront in 1846, one of the largest buildings in the city at that time.

By 1877, the Iron Works employed 600 persons and used 40 tons of scrap and pig iron per day. Thirty-two thousand tons of iron were used annually in the production of nails, hoops, rods, castings, and other iron products required in the construction, shipping, whaling, and textile industries. The Iron Works produced 115,000 kegs of nails annually used in the construction of the northeast's rapidly growing cities. (25-45)

The group of Fall River entrepreneurs that controlled the Iron Works became known as the "Iron Works Group" and were instrumental in starting many other enterprises in the city. This group was led by the dynamic Richard Borden, who assumed effective control of the Iron Works following the passing of Major Bradford Durfee, who died from exertion on the day following the great fire of 1843. The Iron Works Group initiated and controlled steamboat service from Fall River to Providence and from Fall River to New York City, the Fall River Railroad connection to Boston, the Cape Cod Railroad, the Fall River Gas Works, several cotton mills, the American Print Works, a textile machinery company, banks and various waterfront developments. By the 1850's, the Iron Works Group was the dominant interest in the city, due to the greater scope and variety of its operations made possible by outstanding management skill, ownership of waterfront property, and access to water power on the Quequechan. (73-29)

The resourcefulness of the Iron Works management was remarkable. When it needed new facilities or capacities, it fashioned them in such a manner as others would take advantage of them. The gas works, the steamship line and the railroad were examples of this strategy. Also, when it needed capacity, it provided it for itself. When it needed additional machinery for its nail factory, for example, the company built the machinery on its site. The company operated its own cooper's shop to manufacture the thousands of nail kegs and barrels that it needed for its finished products. (32-64)

When Providence became a major market for its products, the Iron Works purchased a site on the Providence shoreline and built a wharf, an office building, and a storehouse there. Under the dynamic leadership of Holder Borden, the Iron Works established the Fall

Print Patterns: *American Print Works, Fall River, Mass.*

fig. 11.4 **A sample of print patterns from the American Print Works**

From a publication of the International Exhibition, 1876, also known as the Centennial Exhibition, in Philadelphia. The caption in the publication reads, in part: "These figures may fairly be taken to represent the fashion of the day ... neat, carefully designated, with proper regard for the color-effect, and although pleasing and attractive to the eye,"

Graphic courtesy of the Fall River Historical Society

River and Providence Steamboat line in 1827 to carry its products to Providence, which subsequently became a passenger service providing daily regular and punctual travel between both cities. The owners of the Iron Works founded what later became the Fall River Steamship Line, providing daily freight and passenger service between Fall River, Newport, and New York City. (32-64)

In order to assure that the Iron Works had sufficient coal for its foundries and to power its machines, Richard Borden formed the Borden Mining Company and in 1849 purchased a large tract of land in Frostburg, Maryland. He developed and expanded a coal mine there and sold the coal not only to the Iron Works but also on the open market. The mining business was continued by Richard's brother Jefferson. (25-43) Before that time, expensive bituminous coal was shipped from Pictou, Nova Scotia. (32-65)

After 1880, competition from new steel factories located near to the sources of ore and coal in the Appalachian Mountains and the mid-west made the manufacture of iron products in New England uneconomical, and the Iron Works ceased operations. The new owner, M.C.D. Borden, tore down the Iron Works buildings and focused the company on the growing business of printing cloth. (32-63)

The American Printing Company

Fall River's textile industry specialized overwhelmingly in the conversion of cotton into print cloth. Virtually all of the city's mills produced this "gray" cloth. For the cloth to be marketable, however, it had to undergo bleaching to white cloth and then dyed in solid colors or printed in patterns.

Within a short time, however, the amount of cloth being produced on the Quequechan River could not be accommodated by Andrew Robeson's print mill. To meet this demand, Holder Borden and stockholders of the Fall River Iron Works organized the American Print Works in 1834, the predecessor of the American Printing Company. The American Print Works began operations in January, 1835 on the lower falls of the Quequechan River. By 1840, the Print Works had doubled its capacity and, following the Civil War, built a large new building during the first post-war expansion. Its capacity was now 80 million yards a year, making it the largest print works in the country. (81-228)

The American Print Works was the most prominent legacy of Holder Borden to Fall River and the business world. Holder Borden was born in 1799 and his late teens began working for David Anthony in his Fall River Manufactory. After a few years, he moved to Pawtucket to work for the Wilkinson brothers, in-laws of Samuel Slater. From there, he became the agent for the Blackstone Company, owned by Brown and Ives. Henry Earl described the young and daring Borden as a "thorough business man, a merchant as well as a manufacturer, knew how to buy and how to sell, varied his productions to suit the market, gave up old methods when new ones were better, and so kept fully up to, if not a little ahead of the spirit of his time." (25-35)

A man of seemingly boundless energy, Holder Borden convinced Nicholas Brown and

fig. 11.5 **The American Print Works and the Fall River Iron Works**

Graphic courtesy of Marc Belanger

Moses Brown Ives of Providence to invest in the Massasoit Mill in 1830 and to manufacture cotton goods there under the name of Brown, Ives, and Borden. He subsequently purchased ownership rights in the Troy Mill, the Fall River Manufactory, and the Fall River Iron Works, owned by his uncles, Richard and Jefferson Borden, who were his contemporaries in age. After a year as agent of the print works at the Globe, he became the prime mover and manager of the American Print Works, which was affiliated with the Iron Works. In 1828, he convinced Iron Works shareholders to purchase a steamboat and to begin regular service between the Iron Works wharves and Providence.

While he was immersed in several activities in Fall River, he was still the agent for Brown and Ives' Blackstone Company in Providence. He was a lover of good cigars and horses. He had his cigars made to a length that they would last on his trips between Fall River and Providence. Fresh horses had to be kept at outlying towns to keep up with his pace of traveling the 18 miles between the two communities in an hour. "Farmers said he went by like a streak of lightening, while young boys sometimes gathered on the road to watch for him. The glowing point of those long cigars, like a tiny meteor in the night, marked his dashes between the two towns." (10-29)

During the financial panic of 1837, Holder Borden, sick on his deathbed with tuberculosis, still summoned his nephew to ride swiftly to Providence in a blinding rainstorm to pledge his holdings in order to save the local bank. Within two years of the start-up of the American Print Works, Holder Borden died at the young age of 38.

Jefferson Borden then returned from Providence to assume management of the company, which thrived under his direction for 39 years, until 1876. When the Print Works obtained a charter of incorporation in 1857, Richard Borden became its president and continued in that role until his death in 1874. (25-37)

In 1857, the American Print Works took control of the Bay State Print Works at Globe Corners. The capacity of the two plants together amounted to 100 million yards a year, an amount, however, only equal to a third of the city's production of 300 million yards annually of gray print cloth. The ability of the print works to expand depended on the availability of finishing capacity of local bleacheries. That capacity was considerably increased in 1872, when the Fall River Bleachery began operations and took advantage of the pure water available from Sucker Brook at South Watuppa Pond.

Until 1876, all of the considerable expansion of the Print Works was funded internally, with no capital needed outside of the Iron Works. (29a-557) In 1880, the elderly Jefferson Borden stepped down as the head of both the Fall River Iron Works and the American Print Works and the business was reorganized. The Fall River Iron Works operated the company's cotton mills, and the American Printing Company managed the bleaching, dying, and printing of the finished fabric.

Leadership of the business passed down to two sons of Richard Borden, Thomas and Matthew. Thomas J. Borden had been the manager of the Bay State Print Works in the Globe before it was purchased by the American Print Works. In this new arrangement, Thomas operated all of the Fall River manufacturing business and his brother, Matthew Chaloner

Durfee (M.C.D.) Borden, directed the sale of goods from an office in New York City. (75-18)

Both brothers continued to operate the business jointly until 1886, when Thomas sold his shares to M.C.D. Borden, who became sole owner of both the Iron Works and the American Printing Company. (75-18) The new owner promptly closed the iron-making facilities of the Iron Works, since it could no longer compete with mid-western iron factories, which were closer to both iron ore and coal.

After a few years operating the American Printing Company from his New York office, M.C.D. Borden became increasing impatient with the suppliers of his print cloth, the Fall River textile mills. He came to believe that "he was sometimes trifled with by the treasurers and agents of the mills whose cloth it was his business to print." (75-31)

fig. 11.6 **M.C.D. Borden starting the Corliss steam engine in Mill No. 4, American Printing Company, 1895**

Source: Speed: *A Fall River Incident*

In *The Spindle City: Labor, Politics and Religion in Fall River, Massachusetts*, Philip T. Silvia observes that:

> Borden was a rugged individualist, a fierce competitor who became the bane of many a Fall River mill treasurer's existence. The seeming pleasure that he derived, beginning in the 1890's, from placing Fall River treasurers in discomforting situations sprang from a spirit of revenge for past wrongs. Before the Iron Works [textile] mills were built, manufacturers had often refused to sell cloth for his printing company until they had squeezed unreasonably high prices out of its owner. (72-526 and 527)

Not someone to be "trifled with," M.C.D. Borden decided to supply his own cloth for his enormous printing business. He began a hugely ambitious building campaign, tearing down the remaining Iron Works buildings to build a complex of cotton mills. Mill No. 1 was built in 1889, followed by Mill No. 2 in 1892, followed by Mill No. 3 in 1893, and, the largest, Mill No. 4, in 1895. What was to have taken over 10 years was accomplished in only six. He finally completed his building campaign with Mill No. 7 in 1905-06.

In 1895, John Gilmer Speed noted that:

> He determined to have cloths of his own to print, and so, in a measure, be independent, while at the same time reducing the cost of production ... Mr. Borden had in his mind at the time a plant which ultimately should produce a considerable portion of the whole output of his print works. (75-31)

And so he did. Not satisfied with achieving independence from the Fall River mill treasurers, M.C.D. Borden invited the world to witness it. He commissioned a special voyage of the Fall River Line's steamship *Priscilla* and filled it with New York notables who arrived in Fall River on October 17, 1895, to observe the start of the huge 3,000 horsepower Corliss engine in Mill No. 4, the largest in the United States and custom-made in the Bethlehem Iron Works. Tours were given of the print works facilities and a testimonial luncheon was served on the *Priscilla*. The many speeches included praise heaped on Mr. Borden for his business acumen. At the end of the event, Mr. Borden announced that he would donate $100,000 to Fall River charities, a considerable sum in 1895. (75-92)

To memorialize the event for current and future generations, Mr. Borden commissioned a book on the event, complete with etchings, which described in considerable detail his amazing feat and the day's activities and speeches. This book, by John Gilmer Speed, was titled *A Fall River Incident, or, a small visit to a big mill* and was published in 1895. Matthew Borden seems to have been determined that no one forget this event and his impressive accomplishment.

As a principled manager, Borden had previously earned the esteem and affection of his textile workers, but "from this time forth," says Philip T. Silvia, "Spindle City mill operatives began a love affair unlike any ever known with this mill owner." Distinct from

Matthew Chaloner Durfee Borden

M.C.D. Borden (1842-1912), the third son of Colonel Richard Borden, was educated at Andover and Yale. He was a connoisseur of fine books, paintings, yachts, and horses. When the government leased his steam yacht during the Spanish American War to carry dispatches from Key West, he promptly proceeded to buy a better one. While he spent most of his time in New York, and was a director of many New York banks and corporations, his influence in Fall River was considerable. Known for his shrewdness in manipulating cloth markets to the competitive advantage of his printing mills, he wished to be known as the "grand mogul" of print works. He wrote: "I believe in success, and the greater the better. I believe in the accumulation of wealth, without any limit, except, always, that fixed by clean and honorable methods, but I believe, also, that unusual success brings with it inseparably extraordinary responsibility." (24-108)

fig. 11.7 **M.C.D. Borden**

other Fall River mill owners, "he was a new type of manufacturer, the soft line entrepreneur concerned about his employees and annoyed by his peers who had always paid slavish wages." (72-527 and 528)

In a turn of the century *Cosmopolitan* magazine article, M.C.D. Borden explains his philosophy of business:

> Wages must be kept up, because the cheapening of manufactures cheapens prices. I am a believer in good prices, and believe that the men are entitled to share in the benefits. This business is not a secret, but the road to success is open to all. If competitors cannot pay as high as I do, either they are too selfish to share the profits with their men, or they do not know their business and have not succeeded in making it pay. (72-527)

M.C.D. Borden had brought the American Printing Company to its pinnacle of success. That company printed three million yards of cloth a year, surpassing in volume any other print works in America or Europe.

Shipbuilding Activity of the Iron Works

Since the Fall River Iron Works was founded by two shipbuilders—Major Bradford Durfee and Colonel Richard Borden—it was logical that the founders of the firm would continue to build sailing and steam vessels. Before the founding of the Iron Works in 1821,

Major Durfee and Colonel Borden built one vessel a year from their shipyard at the mouth of the Quequechan River. (25-47)

While the first steam-powered vessel was developed in 1794, steam power did not come into general use until about 1828. Sailing vessels carried major freight to and from Fall River for another 110 years. The last sailing vessel to service the city was the Somerset-built schooner *L.L. Simmons*, which made its last run from Fall River in 1936.

Somerset was the leader in shipbuilding along the Taunton River during the days of sail. There was little shipbuilding activity in Fall River except for that of the Iron Works. In 1858, Joseph C. Terry built the *Weetamoe*, a steam ferry commissioned by the Brightman and Slade families for carrying passengers and freight across the Taunton River at Slade's Ferry. Terry's shipyard was located on the Taunton River just north of the outlet of the Quequechan River.

Before owning the Fall River shipyard, Terry worked with the renowned James Hood at his Somerset Village shipyard on the Taunton River. Hood and Terry built some of the fastest and most famous of the clipper ships of that era, including the *Raven* and the *Governor Morton*.

Because of the Iron Works, Fall River became a significant shipbuilding site. The Iron Works built its own ships to bring raw material to its foundries (coal, iron, iron ore, and other materials) and to carry its finished products to various markets. These sailing and steam vessels had a freight carrying capacity ranging from 100 to 400 tons.

They included the sloops *Ann B. Holmes* and *Isaac H. Borden* and the schooners *Sea Bird, Minerva, Richard Borden, Ellen Barnes, Jane F. Durfee, Iram Smith, Enoch Pratt, Daniel Brown, Sallie Smith, Orion, Saphronia, Anna M. Edwards, Ney, Martha Wrightington, Thomas Borden, Matthew C. Durfee, Carleton Jayne,* and *Fountain.* (32-65)

Even before the advent of the railroad in Fall River, the Iron Works built a marine railway in 1833 and operated it in conjunction with the shipyard. The marine railway was the first of its kind on the Taunton River and Mount Hope Bay. (29a-557) Steamships and sailing vessels were brought up the marine railway to the shipyard for repairs. The Iron Works also manufactured boilers and other machinery for steamboats.

The last vessels were built in the 1870s, in the next decade, when the iron manufacturing business of the Iron Works ceased, the company focused solely on making cotton print cloth. These last ships were the schooners *D. M. Anthony* and *Carrie S. Hart,* and the barkentine *David A. Brayton*, all of about 800 tons capacity.

The Iron Works and its Local Steamboat Service

Before the Fall River Iron Works initiated its steamboat service, the only way to travel to other locations was by horse and carriage or by sail.

During the early days of passenger travel by sail from Fall River, trips were unscheduled. If a captain was sailing to Providence, Bristol, Newport, or beyond to New York, Baltimore and Philadelphia, he would advertise the trip by word of mouth or placard several days in

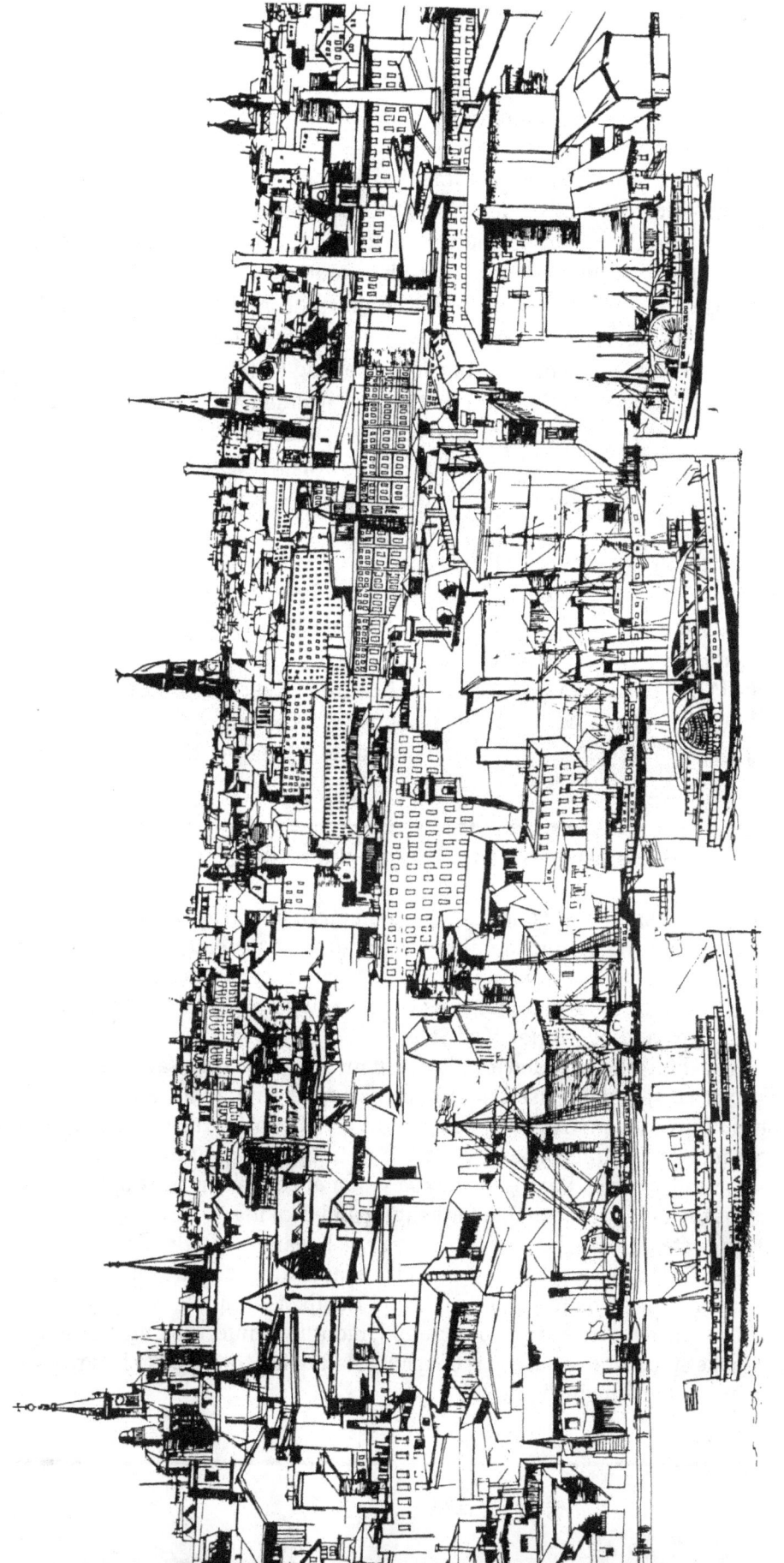

fig. 11.8 **Fall River's waterfront, about 1900**
Source: David Symons in *Fall River 2000*

advance, giving the day of departure and the number of persons that could be accommodated. Some captains began to cater to this trade with attractive cabins and excellent meals. At first, this trade was centered in Somerset, but as Fall River began to grow as a textile center after 1813, it became the focus of passenger travel by water.

As this passenger business became more important and profitable, regularly-scheduled passenger packets became the norm, with vessels built for speed and passenger convenience. Regularly scheduled vessels from Somerset and Fall River left for Providence, Bristol, Warren, New York, Philadelphia, Baltimore, and Savannah. William Lawton's packet Industry from Somerset ran as far as Havana on regular schedules. The trip from Fall River to New York on a sailing packet took 24 hours one-way, with favorable weather. The packet left at 8:00 AM and the fare cost $2.00 to $5.00, depending on accommodations. Passengers slept on board. (41-86)

It is noteworthy that ships from Fall River always went south, not north to Boston and other northern ports. This is because the trip north would require rounding Cape Cod on rough, open ocean, with vessels vulnerable to shifting shoals and frequent fog. It was a long, hazardous trip. The trip south to New York, on the other hand, was by the protected inland waterway of Long Island Sound. A trip from Boston to New York required taking the stagecoach from Boston to Fall River and then sailing to New York.

Early freight hauling by sail at Fall River occurred in vessels that sailed between

---

Steamboats on the Quequechan River

While the Quequechan River is not usually thought of as navigable, for over 200 years the river was the site of considerable water-borne commerce. In the Colonial era and later, this shipping activity was in sailboats, which carried passengers and freight from the shores of the Watuppa Ponds to the growing village of Fall River. The destination was "The Landing" at Hartwell Street.

Most of the freight traffic included logs to be cut into lumber at the saw mills on the Quequechan River falls, tanbark used in the processing of animal hides into leather, sand and gravel for use at construction sites (including mortar for the construction of mills), and voluminous amounts of cord wood for use in the village and in the towns of Newport, Bristol, and Warren. Before the introduction of coal to the area in the early 1800s, wood was the only fuel available for heating and cooking.

In the 1840s, a steam vessel called the *Enterprise* began plying the waters of the Quequechan River. It also carried passengers and the same freight as the sailing vessels but also added day excursions to picnic groves on North Watuppa Pond. In order to sail under the fixed bridges at the Narrows and Plymouth Avenue, the *Enterprise* was fitted with a smoke stack whose base was on a hinge, allowing the stack to be lowered when approaching one of the bridges.

Providence and Taunton. The freight was loaded on and off at Slade's Ferry, since there was no adequate wharfage at Fall River village. Cotton and other merchandise was loaded here for many years. The first regular water communication between Providence and Fall River was by a small schooner, which was large enough to hold ten bales of cotton and a cargo of flour and other goods. This was succeeded by the sloop *Fall River*, which was succeeded by the sloop *Argonaut* and another craft. Water freight service between Providence and Fall River continued in sailing vessels until steamboat service was initiated. (25-10)

The first recorded steam boat in the United States was built by Captain Samuel Morey. It sailed down the Connecticut River and into Long Island Sound in 1794, as George Washington was beginning his second term. In 1807, Robert Fulton sailed the *Clermont* down the Hudson River. However, the loud, unreliable and smoky steamboats were not an overnight success. The early steamers burned so much wood that the stacks of cordwood left little room for freight. Occasionally, a sailing vessel would have to restock the steamers with wood in mid-trip.

The first regularly scheduled steamboat service on Mount Hope Bay was established by the Fall River Iron Works in 1827. The company purchased the steamer *Hancock* and initiated the first steamboat service between Fall River and Providence in that year. The Hancock also ran on alternate days from Fall River to Newport. (41-87)

In 1832, the *Hancock* was superseded by the *King Philip*. In 1845, the Bradford Durfee provided service to Providence and was followed by the *Metacomet* in 1847 and the *Canonicus* in 1849. The *Richard Borden* was launched in 1874. (29a-557)

In addition to this passenger and freight service to Providence and Newport, the Iron Works also operated its own large fleet of sailing and steam vessels that carried freight to and from the company's wharves in Fall River.

While steam was gaining ascendancy, the captains of the sailing packets were not about to be put out of business that easily, and for years they handily out-sailed the crude early steamers. Faster sailing boats were built on the Taunton River to compete with the steamers and the battle between sail and steam waged for 20 years. Combination sail and steamboats were an intermediary hybrid. Finally, the increasingly reliable steamboats could not be beaten on the longer runs, and Taunton River sailing vessels turned exclusively to freighting. (41-87)

The Iron Works and its Railroads

Before 1845, the only way to travel to Boston overland from Fall River was by stagecoach or on horseback. For example, Jesse Eddy, the buyer and seller for the J & J Eddy satinet mill on the Quequechan River, had to travel all over New England and some areas of the West on stage or on horseback to purchase the necessary supplies of wool for the factory. In addition, the business required a weekly trip to Boston that he made in his own carriage (there was no daily service), leaving one day and returning the next. On some days, when dispatch was necessary, he departed and returned on the same day, using a relay of horses provided at inns along the way. (25-32)

Beginning in 1835, access to Boston by the new and novel train service was available at Providence (via the stage or on horseback from Fall River), and access to New York via the new steamboats could be had at Stonington, Connecticut. (25-50) The Fall River to Stonington trip was by sailboat.

The first railroad servicing Fall River began passenger service on June 9, 1845. It was established by Colonel Richard Borden and was named the Fall River Railroad. It provided service from Fall River to the Myricks in Berkley. From there, connections were available to the Taunton-New Bedford branch of the Boston and Providence Railroad. The track from Fall River was later extended to Braintree, where more direct lines to Boston were available via the Old Colony Railroad. This new route reduced travel time between Boston and Fall River. (25-50)

With a view to making the Fall River Railroad self-sustaining, Colonel Borden organized the Cape Cod Railroad, from Middleborough to the Cape, as a feeder line for his Fall River Railroad and its steamboat service to New York City. (25-50)

The Fall River Railroad's connection to Braintree became profitable to the Old Colony Railroad because of the increase in ridership resulting from the connection to Fall River's New York steamboat service. In 1847, managers of the Fall River and Old Colony railroads agreed to a division of revenues. However, in 1852, the Fall River interests had come to believe that the Old Colony was getting the better of the deal. The dispute was resolved in 1854, when the Bordens were bought out of the railroad business and the two railroads came together as the Old Colony and Fall River Railroad.

Old Colony interests in Rhode Island then wanted to build a connection between Fall River and Newport. However, that would have meant that the Borden's profitable steamboat business to New York City would be ruined by a Newport rail terminus. Why stop in Fall River when the boat train could proceed to Newport?

> Alarmed by the prospect of Fall River becoming a way station, and not too fond of the Old Colony managers in the first place, the Bordens did all they could to hinder the Old Colony's efforts to build through Fall River. The Newport line was finished in 1862, at considerably greater expense than the Old Colony had at first estimated. For one thing, the Old Colony Railroad now owned a steamboat line, having bought out the Bordens in a liberal settlement of their disputes. (37-117)

Fall River's Bay State Steamboat Company then became Old Colony's Boston, Newport, and New York Steamboat Company. The Old Colony board of directors then voted to immediately begin using Newport as the terminus of its new steamboat company. Soon after, the Old Colony built a new rail line via Taunton, crossing the Taunton River at the Fall River-Freetown border. That new line reduced the running time of the boat train by 15 minutes. (37-117)

A RIVER AND ITS CITY

More direct rail service from Fall River to Providence became available beginning in 1865 via a ferry from the bottom of Ferry Street. This ferry transported passengers from the Old Colony RR terminal on the Fall River shoreline to the terminus of a rail line in South Somerset that ran to Providence via Warren. This ferry service continued until December 5, 1875, when the Slade's Ferry Bridge opened and provided rail service across the lower part of the Taunton River.

Because of the 130-foot difference in elevation between the new railroad at sea level and the top of the hill, a direct rail connection between Fall River and New Bedford was impossible. In 1875, direct rail service between New Bedford and Fall River was established by bringing the line from New Bedford and ending it at Plymouth Avenue, near Wordell Street. By this time, the city was too developed on either side of the Quequechan River to bring a rail line through it. Given that the only place for a railroad to approach Fall River was through the Narrows and along the river, the rail line was built in the middle of the Quequechan River on a causeway and on pilings. This Watuppa Branch of the Old Colony RR had a station at its terminus at Plymouth Avenue.

fig. 11.9 **The Fall River Line steamer *Bristol***

Source: Earl: *The Centennial History of Fall River, Massachusetts*

In 1900, bold plans were developed to make a direct connection between the station at the Fall River Line Pier and the Watuppa station via a tunnel cut into the granite base of the city. Rights to land were purchased to begin the project, but it was eventually abandoned. This connection would have allowed a direct rail link between the Fall River Line and Cape Cod.

The Fall River Line

The Iron Works early operation of the Fall River to Newport steamship service—and its connections with the Boston stagecoach—led inevitably to an interest among the Bordens in establishing a Boston to New York rail/steamship route once the railroad was extended to Fall River.

The origins—and success—of the Fall River Line are in geography. Continuous rail service between Boston and New York was possible as early as 1835. However, it was an uncomfortable, long trip with poor sleeping accommodations. The extremely wide mouth of the Thames River in Connecticut provided a formidable barrier for a rail crossing along the coastline and indeed was not crossed with a rail bridge until almost 1890. The railroad therefore had to travel inland to cross where the Thames was narrower. It was possible to take the train from Boston to Stonington, north of the Thames, then board a steamboat to New York City. However, that required a transfer in mid-trip, and the Stonington service was soon eclipsed by Fall River steamers.

The most significant geographic factor that favored Fall River was the peninsula of Cape Cod, which discouraged a direct water connection between Boston and New York City.

fig. 11.10 **The New Line boat, Fall River**

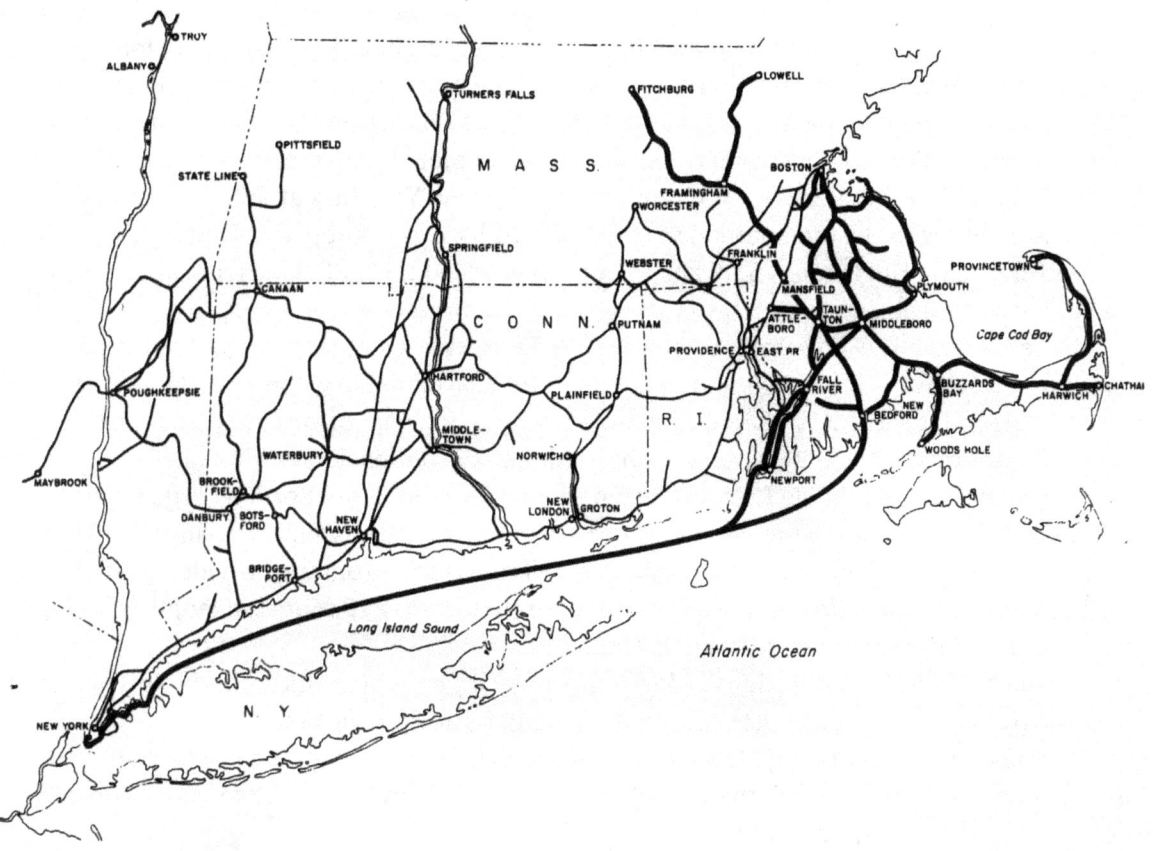

fig. 11.11 **Rail lines serving the Fall River Line steamers**

Fall River's location south of Cape Cod gave the Fall River Line a strategic advantage, since it allowed a short rail travel time once the steamboat portion of the journey was complete, allowing passengers to sleep later in the morning. The rail linkages leading to Fall River allowed the service area of the Fall River Line to span most of Eastern Massachusetts, from Fitchburg and Lowell to Provincetown.

Reproduced from Foster: *Splendor Sailed the Sound*

The Fall River Line route was 228 miles, almost all of it on the inland waterway of Long Island Sound, coupled with a short rail trip to Boston. The trip around the Cape, however, was 337 miles and was over rough open waters that were subject to frequent and dangerous fogs, shoals, and storms.

The Fall River Line had another advantage. Of all the Long Island Sound boat-rail routes, the Fall River service had the shortest rail travel time once the steamboat portion of the journey was complete. This meant that passengers on the Fall River Line could remain asleep later in the morning. Still another advantage was that the Boat Train was waiting right at the wharf. (37-111)

The length of the Fall River line trip lent itself to boarding the train in Boston at the end of the workday, getting on the boat at Fall River an hour later at dinnertime and, following a comfortable night's sleep, arriving in New York in the morning at the start of the business day. Boston passengers going to New York boarded the Boat Train at 6:00 pm and arrived in Fall River at 7:26 pm. The steamers left the Fall River pier at 7:30 pm, took on passengers at Newport, and left that city at 8:30 pm, arriving in New York City at 7:00 am. New York passengers going to Boston boarded the steamers at 5:00 pm, stopped at Newport at 3:00 am and arrived in Fall River at 5:00 am. They then boarded the 5:20 Boat Train and arrived in Boston at 6:50 am. (37-118)

Once the new rail line from Boston to Fall River was competed in 1845, steamship service to New York City began immediately on the leased *Eudora*, the first steamboat to operate the Fall River to New York City route. During its first year of operation, the relatively small *Eudora* carried 50,000 passengers between the two cities.

The route was profitable from the beginning and, in 1846, the Bordens organized the Bay State Steamboat Company. A new boat was ordered for the nascent company and called the *Bay State*. Another boat, the *Massachusetts*, was chartered from the Providence line. On May 19, 1847, the *Bay State* made her first trip to New York. At 315 feet in length, the *Bay State* was the largest inland steamboat in America. (37-112)

As business grew, the Bordens ordered the new *Empire State*, which arrived in 1848 to replace the leased *Massachusetts* and, in 1849, bought the *State of Maine*.

In 1854, the 342 foot-long *Metropolis*, a radical departure in marine construction, went into service. On June 9, 1855, the *Metropolis* sailed from New York City to Fall River in 8 hours and 21 minutes, a record that stood for 52 years. (37-113)

The Fall River Line steamers grew larger and more lavish. The *Metropolis* was followed by the *Newport* (1865), her sister ship the *Old Colony* (1865), the *Bristol* (1867), the *Providence* (1867), and the *Pilgrim* (1883).

The *Pilgrim* was designed by the legendary George Pierce, who revolutionized steamboat design. The *Pilgrim* was the first American vessel of iron construction to have a double-hull, with watertight compartments to improve flotation. When she hit an uncharted rock in 1884 and opened a 100-foot gash in her hull, she was able to steam safely to port unaided. No other boat at that time could have experienced such a large opening and remained afloat. Her main attraction, aside from her size and elegance, however, was that the *Pilgrim* was equipped entirely with Mr. Edison's new electric lights. (37-130)

The *Puritan*, 419 feet in length, began service on June 17, 1889, and established a new standard for elegance. Its interior was designed in the Italian Renaissance style with gold inlaid detailing. The *Plymouth* was launched in 1890, followed by the *Priscilla* in 1894, which was then the world's largest sidewheeler at 440 feet. In 1905, the *Providence* was launched and, in 1908, the elegant *Commonwealth*. (37-150)

The *Commonwealth* was fondly called the "Queen of the Sound" because of her enormous size and beauty. She carried 2,000 passengers and was essentially a lavish six-story floating hotel. While she was shorter than the ocean-going *Lusitania* of the Cunard

fig. 11.12 **The Fall River Line steamer Commonwealth**

The *Commonwealth*, fondly called the "Queen of the Sound" because of her enormous size and beauty, was launched in 1908 and was the last of the Fall River Line's great legacy of steamboats plying Long Island Sound to New York City. She was the size of the largest ocean liners and carried 2,000 passengers. She was essentially a lavish six-story floating hotel.

Source: David Symons in *Fall River 2000*

Line, the *Commonwealth* was wider and could carry almost as many passengers and just as much freight. (37-297)

While consideration was given to constructing the line's later steamers with propeller screws, it was decided to keep the side-wheel design on all of its steamers up to and including the *Commonwealth*. "The side-wheelers were able to deaden headway and reverse very quickly, a factor of no mean importance on a body of water where fog was frequent and marine traffic heavy." (59-91) The paddlers were more maneuverable, particularly in reverse, which in part accounted for the Fall River Line's excellent safety record. Many of the line's captains could dock these huge steamers without the benefit of a tugboat. The shallow approach to Fall River along Mount Hope Bay was also a consideration in keeping the shallow draft side-wheelers. As one ship manufacturer is reported as saying, "The Fall River Line wants a four-story hotel with shallow draft, and you can't do that with a propeller." (37-136) In addition, the great beam of the Fall River Line steamers gave an interior spaciousness to them that was notably lacking in their propeller competitors. (59-91)

While the Fall River Line was best known as a passenger line, it was also a freighter service. The steamer *Plymouth*, for example, could carry the equivalent of 72 railroad boxcars. (37-147) This daily freight service to New York City provided enormous advantages to Fall River as a textile center. An order for cloth from New York to Fall River could be placed on an afternoon and be delivered at the foot of the garment district the next morning. Major inland textile centers such as Manchester, Lowell, and Lawrence had no such advantage. (73-63)

Other freight carried by the Fall River Line included fish from Newport, once the largest fishing port in New England. For example, on May 20, 1887, a total of 1,400 barrels of scup were loaded on the Fall River steamers bound for New York. Boats often had to be held an hour or more beyond the Newport sailing time to load fish. The line also carried freight from communities along the Taunton River, such as pottery from Somerset and strawberries and flowers from Dighton.

When Richard and Jefferson Borden sold the Bay State Steamboat Company to the Old Colony Railroad in 1862, the new owners transferred the terminus of the boat train—and its steamship service—to Newport. In 1867, Jim Fisk's Narragansett Steamship Company opened a competing steamboat service in Bristol with the steamers *Bristol* and *Providence*. Both of these ships were clearly superior to anything else on Long Island Sound in terms of size and elegance, and they began to drain business away from the Old Colony boats at Newport. The Newport terminus had another problem: because it was further south from Bristol, its earlier docking time of 3:00 or 4:00 in the morning meant that passengers could not sleep as long as on the Bristol trip. (37-119)

This constraint, and competition from other Sound steamship lines, forced the Old Colony Railroad to sell its steamship business to Fisk's Narragansett line. As part of the agreement, however, Old Colony insisted that the Bristol connection be abandoned so that it could continue to have the steamship line's rail business. Another reason for moving out of Bristol was that its harbor was a relatively hazardous place for the larger steamships that were now being built for the Long Island Sound business. (37-119)

Fisk then moved the terminus of the steamship line back to Fall River in June 1869, where it remained. As Fisk and Jay Gould, his business partner, arrived on the Bristol that June morning, Fall River gave the two men and their entourage a rousing reception as the city welcomed the line back to the Spindle City.

Fisk was an ideal owner of a steamship line. He was an excellent promoter of the reconstituted Fall River Line and improved the service and elegant appointments on the steamships until they acquired the names of "floating palaces." He upgraded the kitchen and dining experience until it compared favorably to the best class hotels anywhere. He added to the excitement of steamboat travel by providing band and orchestra concerts every evening. Fisk can be credited with making the Fall River Line "the only way to travel" between New York and New England. (37-120)

fig. 11.13 **The Fall River Line steamer *Providence***

Source: From a painting by Antonio Jacobson at the Peabody/Essex Museum of Salem

fig. 11.14 **The freight steamer *City of Fall River***

Freight steamers were a valuable segment of the Fall River Line service, transporting all manner of freight between Fall River and New York every evening and to other ports along the East Coast. The freight steamers were also used to try out new concepts and technologies. For example, the *City of Fall River* included the first compound engine and the feathering paddle, which were subsequently used in the *Puritan* and later Fall River Line steamers.

Source: Reproduced from Foster: *Splendor Sailed the Sound*

Jim Fisk clearly loved steamboats and delighted in playing "admiral" to his fleet:

> His tailor made him an admiral's uniform including regulation gold stripes and stars but with the Narragansett emblem on its gold buttons and lapels and room for the largest shirtfront diamond stud in the nation. With his plumed hat and resplendent costume, the rotund Commodore Fisk personally presided over the daily debarkations from New York harbor. (10-147)

Fisk's flamboyant style extended to his private life and, in 1872 at the age of 37, he was shot to death in a New York hotel by a jealous rival in a love triangle. Fisk's associate, financier Jay Gould, subsequently became owner of the Fall River Line. The Old Colony board of directors were alarmed at having such a business partner, since Gould had a well-earned reputation as a notorious speculator known for sharp practice and failed enterprises, including the wreckage of the Erie Railroad. (37-124)

The Fall River Line even figured in a Jay Gould plot to corner the gold market. President Grant's brother-in-law was in on the scheme, but for the plot to succeed it was necessary to find out whether or not the Treasury would sell gold to break the corner. Messrs. Gould and Fisk royally entertained the President aboard Providence on a trip to Boston, but Mr. Grant disclosed nothing. Eventually, the Treasury did sell gold, and the boys lost quite a bit of money. (37-121)

For his part, Gould did not share Fisk's fascination for steamboats and, in 1874, sold the Fall River Line to the Old Colony Railroad, which created the Old Colony Steamboat Company to operate the fleet. The new owners maintained the level of service and elegance initiated by Jim Fisk, and business continued to grow.

In 1879, the New Bedford to New York steamship service also came under the control of the Old Colony Railroad. The new owners terminated steamboat passenger service from New Bedford to New York and made it a freight-only service. From that time on, passengers from New Bedford and Cape Cod had to take the newly-built railroad to Fall River to get the Fall River Line boats to New York.

The Fall River Line steamers inspired the imagination. They were the favorite of honeymooners and "were fondly remembered by more than one generation as the scene of some of their most romantic moments." (37-291) In his book *The Old Fall River Line*, Roger Williams McAdam, a Fall River Line enthusiast, rather effusively remarked, "Oh, that everybody in the United States might have seen the *Commonwealth*, a bulk of shining lights, tier upon tier, make her graceful swing around Newport's Goat Island emerald light and point to the open sea! Then she was an effulgence of glory—a fairyland that moved and had being." (59-92) Many of Fall River's older south end residents remember the beautiful and romantic spectacle of the Fall River Line steamers gliding majestically along Mount Hope Bay.

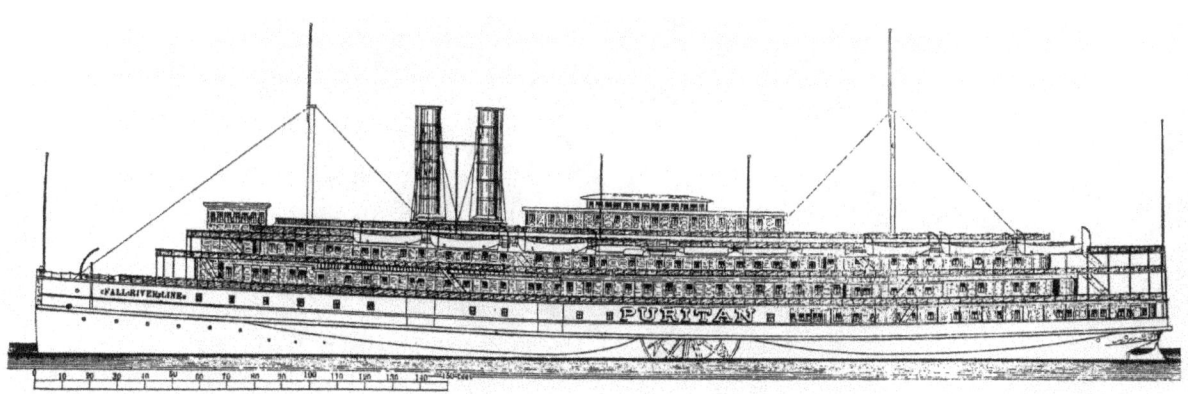

fig. 11.15 **The steamboat *Puritan***

The lavish Fall River steamers appealed to an upscale clientele who were attracted to the line's beautiful interiors and excellent dining, comparable with any first-class large restaurant of the day. Passengers on its "floating palaces" included presidents James Polk, Millard Fillmore, Ulysses S. Grant, Chester Arthur, Grover Cleveland, Benjamin Harrison, Theodore Roosevelt, and Franklin D. Roosevelt. Presidents Hardy, Coolidge, and Hoover were also among the lines passengers, but not while they were in office. (59-71)

Many wealthy New Yorkers were regular passengers to Newport, including the Vanderbilts, Astors, Belmonts, and Rockefellers, among others. Prominent Philadelphia families using the line included the Lippencotts, Clothiers, Whartons, and the Buffams. Pennsylvania coal magnate E.J. Berwind traveled the Fall River Line on his way to his Bristol estate Blythwold. Chief Justice Charles Evans Hughs, General John J. Pershing, and aviatrix Amelia Earhart, among many other notables, were passengers. (59-71)

Of all the passenger steamers plying the east coast, the Fall River Line was special, as Robert G. Albion relates in *New England and the Sea*:

> All the Long Island Sound lines had their share of enthusiastic patrons, but in song, story and fact, there was nothing quite like the Fall River Line. Changes in its corporate ownership were confusing, and its Narragansett terminus moved from Fall River to Newport to Bristol and back to Fall River; but its reputation for regularity, safety, and fine boats was unsurpassed. In the late 1890's, when the New York, New Haven & Hartford Railroad Company consolidated the major Long Island Sound steamboat lines into the corporation that eventually became the New England Steamship Company, all the services were managed by personnel of the Fall River Line. From 1874, when records were first kept, until operations were suspended in 1937, only one passenger lost his life among the nearly 19 million carried. (1-176)

12 The Granite Quarries of Fall River

by Kenneth M. Champlin

During the first half of the nineteenth century, the United States was a nation under construction. Around 1820, Bradford Durfee was getting stone from the Bigberry Ledge in the vicinity of present-day 14th and Pleasant Streets. Durfee would soon achieve the status of a local hero when he devised a way to draw hammered granite blocks upward on the walls of the Anawan Mill in 1825, the same year the "Bunker Hill" Quarry opened in Quincy. Derricks, used for hoisting granite blocks, would not be employed in Fall River's granite quarries until after 1830.

Durfee may have persuaded his nephew, Benjamin Davol, to abandon his ancestral trade of shoemaking and enter into the stonecutting business. By the time he turned 21 in 1822, Benjamin Davol had amassed a fortune of $3,000 and subsequently bought a ledge at the head of Bedford Street later owned by the Beatties. At that time (around 1825) "the head of Bedford Street" lay at the present-day intersection of North Quarry, Bedford, and Quarry Streets. From his quarry, Davol supplied stone for the Groton Monument in Connecticut, as well as for the construction of Fort Adams in Newport. In Fall River, Davol provided "dimension" or building stone for the Market Building and the original Granite Block, two major buildings erected in the town center after the Great Fire of 1843.

In his *Final Report on the Geology of Massachusetts* (1841), Amherst College professor Edward Hitchcock, a geologist for the Commonwealth, extolled the quality of granite found at "Fall River in Troy": "... no rock can be finer for architectural purposes than the granite of Troy, and immense quantities have been obtained from this locality. The large manufactories of Fall River are built of it, as is also Fort Adams at Newport, R.I."

Hitchcock describes the actual quarrying techniques employed in Fall River and other Massachusetts quarries by 1840. One method for extracting granite, "plugs and feathers," was developed at the Quincy Quarry that supplied stone for the Bunker Hill Monument. "Feathers" are the cases of sheet iron or half-round strips of iron, flat on one side for contact with the wedge (or "plug") and curved on the other side to fit the wall of the drill hole. According to Hitchcock, a row of wedges are then driven in at the same time with equal force, thus splitting the stone along a straight line.

When Davol was supplying stone for the Market Building in Fall River, a stonecutter's daily wages had reached a high of $1.60 and a low of $1.47. In general, the average daily wage of all quarry workers (both the skilled stonecutters and unskilled laborers, or "quarrymen") from 1840 to 1880 was $1.70.

Benjamin Davol died in 1861. Davol's executors assumed control of the quarry, which remained in business under his name until 1864. By 1866, the Davol Quarry, along with another site to the north previously owned by William Harrison, was acquired by two brothers, John and William Beattie of Newport Rhode Island. The new firm erected two large sheds for stone dressing, a boarding house, a barn, and a blacksmith shop.

The enterprise employed 70 men, with plans to hire more. The Beatties also had several ongoing contracts for getting out stone. One was a seawall at Castle William, Governor's Island, New York Harbor. Closer to home, orders came for granite to construct a pier in the Mystic River, Connecticut, to repair Long Wharf at Newport, and to build a bank wall on Pocasset Street in Fall River. The Beattie Quarry also had a contract to supply stone to the Old Colony and New Haven Railroad.

John Beattie was considered the senior partner, and he possessed considerable experience in the stone business, especially as a contractor supplying stone to the railroad companies. Beattie gained much of his management experience supervising large work crews at Fort Adams, Newport. After operating the Bedford Street site for only a year, John Beattie left Fall River to open a quarry at Niantic, Connecticut. Eventually, his brother William assumed complete ownership and management of the Fall River quarry.

One noteworthy project in which William Beattie was involved was the construction of the B.M.C. Durfee High School on Rock Street in 1886. An article in the Fall River *Daily Evening News*, June 18, 1887, describes the dedication of the new high school. Included was the full text of Mr. John S. Brayton's address. Brayton mentions that the stone for the first story of the building came from Beattie's Ledge. Stone for the upper stories came from Alexander McDonald's quarry at Mason, New Hampshire. McDonald used a dressing machine of his own invention on the granite required to complete the building.

Coarse-grained "Fall River pink" was used in the first story of the high school. Beattie's granite contrasted subtlety with McDonald's New Hampshire gray granite, a feature of the new building noted by Brayton in his address. Beattie's Ledge also contained a variety of gray granite, "Fall River gray." Also found in other local quarries, "Fall River gray" granite was coarse to medium in texture with large black and greenish spots. Beattie's "Fall River Pink" granite, pinkish gray to tan in color, with specks of mica, was coarse in texture. The directors of the high school project sought another supplier for the stone needed to finish the building.

However, William Beattie did net a contract for $18,480.74 to supply stone for the reconstruction of the old city hall after it had been gutted by fire in 1886. Beattie's Ledge granite was also used in the construction of the Fall River Public Library (1896-1899). When W.L. Rutan, a Boston contractor, was awarded the contract on a low bid of $133,900, the city stipulated that he use only Fall River granite in the project.

In 1868, the lots that contained the Beatties' quarry operations equaled a total of 1,413,151 square feet, or over 32 acres. By 1905, William Beattie's quarry operations were bounded by North Quarry Street on the west and by Beattie Street on the south. In 1910, the quarry itself measured 900 x 700 feet and was 20 x 60 feet deep. According to one U.S. Geological Survey, oxcarts were still used to transport quarried granite a mile and a half west to the railroad or wharf.

fig. 12.1 **Fall River city hall**

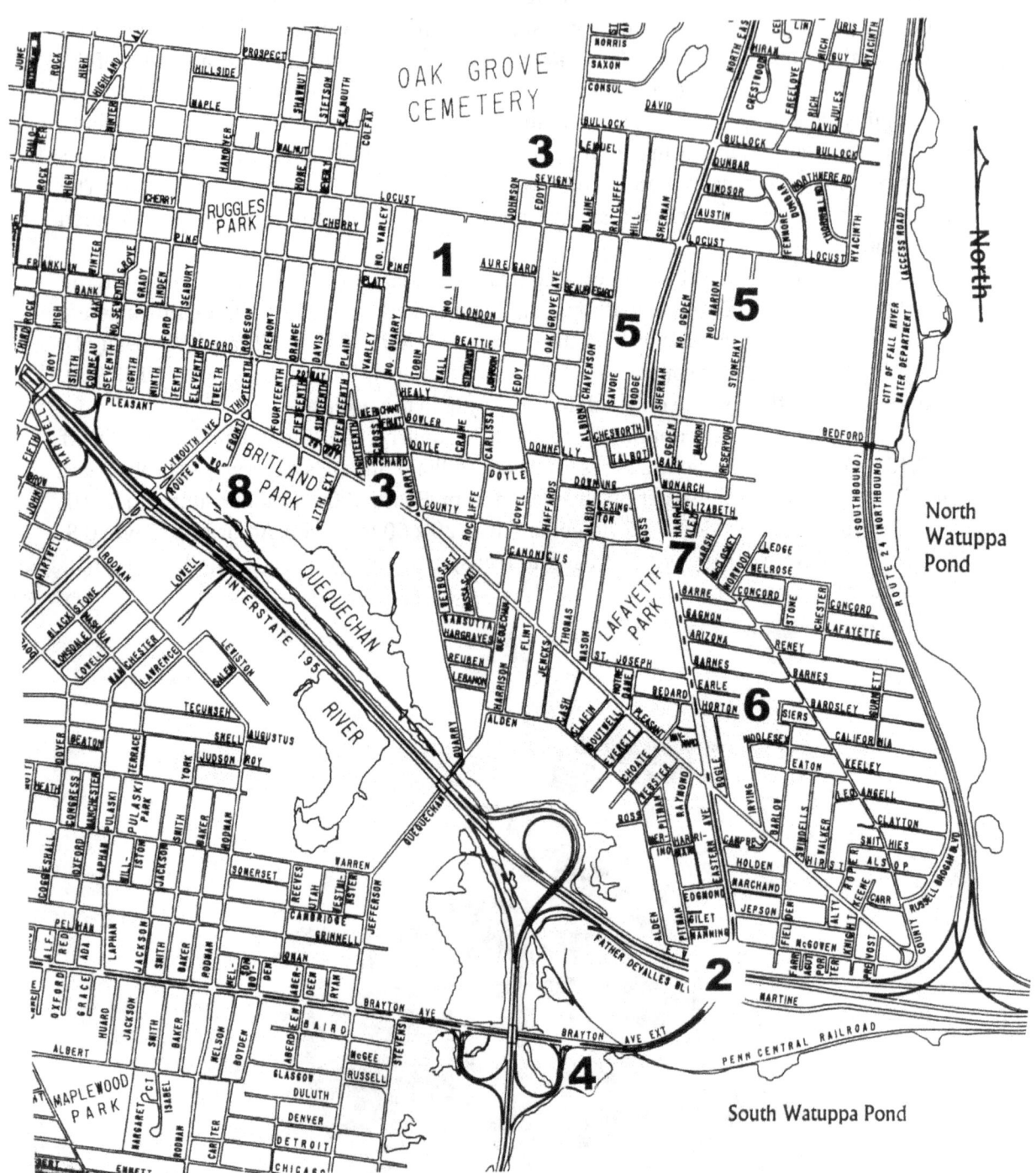

fig. 12.2 **Commercial granite quarry sites in the Quequechan River valley**

Quarry Map Index

1. Originally the Benjamin Davol Quarry, this site was first opened about 1825. In 1866. it was acquired by William and John Beattie. When Davol owned the site, it was described as being "at the head of Bedford Street," since there were no other public roads in existence in the area at that time. Even as late as 1895, North Quarry Street was part of Quarry Street and ended near London Street. By 1905, Beattie's operations, including quarry and outbuildings, were bounded roughly by Beattie, North Quarry, Locust, and Wall Streets.

2. The Beattie and Wilcox Quarry, opened in 1893, comprised of four separate pits at the southern end of Eastern Avenue around McGowan, East Warren, and Kerr (now Martine) Streets.

3. The Stinziano Quarry, opened before 1895, was located in the vicinity of Locust and Oak Grove Avenue, although the actual location was probably at the head of Eddy Street, along Sevigny Street. Pascale Stinziano also probably owned a second quarry at the northwest corner of Quarry and Pleasant Streets, later acquired by Morris Tonkonogy, who had already acquired "Beattie's Ledge."

4. The Chauncy Sears Quarry, opened in 1892, at South Watuppa Pond at the head of the Quequechan River.

5. The Savoie/Carey Quarry (two locations), originally opened as Dudevoir, Savoie and Company quarry in 1895. Later, it was the Savoie Quarry and, finally, the Carey Quarry and Construction Company.

6. The Ross Quarry, opened before 1901, located at the north end of Barlow Street.

7. The Harrison, or Rolling Rock, Quarry. It opened before 1861 and was located where County meets Eastern Avenue at the Rolling Rock.

8. The "Bigberry Ledge" site at what is now Britland Park. This is the earliest site (prior to 1825) where granite was quarried locally. The granite for the nearby Merchants Mill was probably quarried from the area of the Bigberry Ledge.

These are the known commercial granite quarries in the city. Local histories indicate that other smaller quarry sites were opened up for specific projects or limited use.

Alfred J. Lima

The Italians in Fall River

From 1890-1914, Italians accounted for one-third to all stonecutters newly arrived in the United States. Many of Fall River's early Italian settlers were expert stonecutters and masons brought to this country by William Beattie. Italians accounted for most or all of the workforce at Beattie's Ledge, at the Stinziano Quarry (owned by stonecutter Pasquale Stinziano) at the corner of Locust Street and Oak Grove Avenue, as well as at a third, the Beattie and Cornell Quarry located at the Narrows (opened around 1895 by William Beattie's second eldest son, William Henry Beattie, in partnership with George H. Cornell).

Fall River's first Italians established a community, first on upper Bedford Street, then on Healy, Orange, Plain, Johnson, and Quarry Streets. Holy Rosary parish was founded by quarry workers. Romauldo Vioni was a mason and a member of the building committee for the church. He designed a Gothic structure modeled after Notre Dame Church in the Flint. Vioni's plans were considered too extravagant for a small parish; the more modest plans of Ernest Ziroli were substituted instead. Ernest was the architect son of Nicholas Ziroli, a stonecutter for the Beattie and Wilcox Quarry. The Beattie and Wilcox Quarry was opened in 1893 at the southern end of Eastern Avenue by William Beattie's eldest son, David. On June 10, 1904, the unfinished basement church, measuring 115 feet long x 64 feet wide was dedicated. Ten feet of the 14 foot high walls rose above ground and were constructed of granite from the quarry to the north of the church.

The elder William Beattie, the proprietor of Beattie's Ledge, died in 1915. In 1917, ownership of the quarry passed to a group of businessmen headed by Morris Tonkonogy and Barney Prebluda. Some of the outlying land surrounding the quarry was sold off as house lots by the Beattie heirs. However, the Beattie Quarry itself passed into the hands of Morris Tonkonogy intact.

At the time they purchased Beattie's Ledge, Tonkonogy and Prebluda had been operating the former Stinziano Quarry and very likely a smaller pit at the northwest corner of Quarry and Pleasant Streets. Their enterprise was simultaneously known as the "Fall River Granite and Crushed Stone Corp." and the "Fall River Granite and Quarry Co." An ad in the Fall River City Directory touted the firm as providing "cut stone of all descriptions-Foundation stones, building stones, Paving stones, Curbing. Quarry, North Quarry Street, 10 Purchase Street." Up to 1950, the directory listing for the Fall River Granite and Crushed Stone Corp. included the note, "quarry N. Quarry Street."

Morris Tonkonogy specialized in real estate and insurance for 25 years. Known as "the first Jewish realtor in Fall River," Tonkonogy was active in the Jewish life of the Flint. He was founder and treasurer of the Congregation of the Sons of Jacob Synagogue on Quarry Street. He also helped to organize the Talmud Torah Institute on Mason Street. Morris Tonkonogy died May 13, 1939, at his home at 2 County Street, Fall River, at the age of 57.

His surviving partners, Herman Adler and Malvin Wolff, operated the Fall River Granite and Crushed Stone Corp. during the 1940s. In 1951, 12 years after Tonkonogy's death, all mention of the quarry location was dropped from the corporation's directory listing. A

fig. 12.3 **Early photographs of quarrying operations at the Beattie quarry**

As early as 1822, granite was being quarried in Fall River for export. In his *Sketches of Old Bristol*, Charles Thompson mentions that U.S. Senator James D'Wolfe commissioned Bradford Durfee to supply granite for a two-story structure with a "French roof" at State and High Streets in Bristol and "with each block numbered and fitted to its particular place." In 1841, Orin Fowler in his *History of Fall River* mentions that the granite quarries of Fall River were supplying not only the city with stone building materials but was also exporting granite building materials to Newport, New Bedford, New York, Providence, Bristol, and Warren.

Buildings made of Fall River granite included the old City Hall, the old Granite Block, St. Mary's Cathedral, St. Patrick's Church, Norte Dame de Lourdes Church, the Slade School, the lower section of the BMC Durfee High School, the main Fall River Public Library, the Bank Street Armory in Fall River, Fort Adams, and Belford Castle in Newport, the foundation of the New York state capital in Albany and, not least, the many granite textile mill buildings in Fall River.

A tremendous amount of granite was also used for more mundane but essential uses, such as foundations for commercial buildings and dwellings. Concrete did not come into general use for foundations until the invention of the internal combustion engine allowed large batches of concrete to be delivered by truck. In addition to building foundations, granite was used for retaining walls, steps, street curbing, and street paving (cobblestones).

year later, Hillside Manor, a public housing project, was constructed on two old quarry lots, designated in Assessor's Office records as L-1-53 and L-1-54.

In 1955, Aaron Adler, son of the late Herman Adler, and the sole surviving partner Malvin Wolff, donated 9 acres of gradually filled-in land at the head of Pine Street to the city. The gravesite of the 125-year old quarry, upon which stood the housing project, was to be further marked by a playground. The Ralph M. Small School was built on a portion of the donated land.

In 1910, when 3,603 hands were employed in the quarries of the Commonwealth, the United States government undertook a survey of quarry sites in all six New England states. Thirteen years later, T. Nelson Dale used the information gathered by the survey to write U.S. Geological Survey Bulletin 738, "The Commercial Granites of New England." Included in Dale's study are descriptions of six commercial quarries in Fall River. With the exception of Beattie's Ledge (originally opened by Benjamin Davol around 1825), the five other quarries were opened in Fall River's greater East End during the 1890s, primarily in response to the general prosperity of the stone trade in the preceding decade, as well as to meet the demands of the local market.

One of the quarries described in Dale's report, the Sears Quarry, opened in 1892, bordered South Watuppa Pond at the head of the Quequechan River. The owner, Chauncey Sears, built the principal buildings of at least two Fall River mill complexes: the Arkwright (1897) and the Davis mills No. 1 and No. 2 (c. 1902). Also noted by Dale is the Beattie and Wilcox Quarry, opened in 1893 at the intersection of Eastern Avenue, McGowan, East Warren, and Kerr (Martine) Streets, and idle by 1923. The remaining quarries profiled by Dale were the Savoie or Carey Quarry (west of Oak Grove Avenue at the present junction of Savoie and Beaureguard Streets, first listed on Oak Grove Avenue near Locust, opened in 1895, still functioning in 1923), the Stinziano Quarry (also located at the corner of Locust and Oak Grove Avenue, still operating in 1923), and the Ross Quarry (located at the head of Barlow Street, opened before 1901, idle in 1923). The Ross Quarry produced dressed stone for buildings and pedestals, as well as blocks of granite for paving and foundations.

Italian was not the only Romance language spoken in Fall River's granite quarries. West of Oak Grove Avenue, in the area of Savoie and Beauregard Streets, were several small, independent quarry companies, where Canadian French rang out in cadence with hand-wielded hammers. Here, around 1895, Henri Savoie, a stonecutter from Barlow Street, formed a partnership with Joseph M. Dudevoir and James F. Conroy to operate a quarry first listed on "Oak Grove Avenue, near Locust."

The firm was called Dudevoir, Savoie & Co. until 1904, when the name was changed to the Fall River Quarry and Construction Company, located on Beauregard Street. By 1908, Henri or Henry Savoie had withdrawn from active participation in the company. However, in 1910, Savoie become the successor to the Fall River Quarry and Construction Co. That year, Savoie took into his business a recent B.M.C. Durfee High School graduate, 20-year-old John Henry Carey. Carey was the son of Isabella Frances McManus and Edward F. Carey, a Fall River fireman.

When Henry Savoie took control in 1910, the quarry measured 200 feet by 100 feet, and was 30 to 40 feet deep. Savoie granite closely resembled the "gray" of Beattie's Ledge, and was layered horizontally in sheets two to six feet thick. Savoie granite was used in the trimmings of the Westall School and the Fall River Public Library.

Henry Savoie and John H. Carey remained partners in granite and general construction until 1917, when Carey bought out Savoie and changed the name of the business to the Carey Quarry and Construction Company. By 1924, the Carey Quarry and Construction Company operated two granite quarries and, eventually, a third. At its height, the company employed 225 men. According to a Carey descendant interviewed in 1997, one quarry was located in the area "east of Chavenson Street and the south side of Locust." This was the original Savoie Quarry, later expanded to include another area known as Benoit's Ledge. The Careys also ran a second quarry, east of North Eastern Avenue, where Stonehaven Road meets Locust Street. The third and the last of the Carey quarries to close was located at the southern end of Eastern Avenue, near East Warren Street, "where the ramps lead on to 195." This site was formerly the Beattie and Wilcox Quarry, purchased by the Careys from the David Beattie heirs in 1928.

By 1927, the Carey brothers—John, Thomas and Edward—had ceased producing granite on Savoie Street, maintaining only a stable and yard at the old location. They ran the old Beattie and Wilcox Quarry "at the end of Alden Street" until 1932, when the company ceased quarry operations altogether. Again, the company's name was changed to the Carey Construction Company with offices at 56 North Main Street, Fall River, and the Realty Building in Newport. The company kept the old Savoie Street location as its actual base of operation until October, 1967, when the lot was acquired by St. Anthony of the Desert Church.

Omitted from Dale's study was the Harrison Quarry at the base of Rolling Rock. In 1910, the quarry, perhaps idle, still existed in the area of the present-day junction of County Street and Eastern Avenue. In 1861-1862, prior to purchasing the ledge formerly owned by Benjamin Davol (as well as one nearby owned by William Harrison) John Beattie, William Harrison, and his brother Barney Harrison, were getting out stone for the Newport Railroad from the Harrison's small quarry at the base of Rolling Rock in Fall River. Barney Harrison was worried that the Rolling Rock, a 140-ton oscillating "pudding stone," would tumble into the quarry. "I was afraid," Harrison was to have said, "that Alec Borden, Rufe Bassett, Bob Cook and John Mingo would come along and toss the rock over into the quarry while we were at work in it."

While Harrison jokingly made this remark, he was concerned that blasting at the quarry would cause the rock to become dislodged from its perch on the ledge. Barney Harrison decided that the great boulder should be stabilized. To accomplish this, he first drove in some "half rounds" used with wedges to split rocks. Between the half rounds, he drove in fragments of iron wedges, then rammed in stone chips creating a pedestal of iron and stone fragments upon the surface where the rock was balanced. He also placed rocks in an outer ring, "and the wonder of the ages rolled no more."

"Large Blast" was the headline in the Fall River *Daily Evening News* of November 14, 1870, describing the extraction of a massive piece of granite from William H. Harrison's quarry at the base of Rolling Rock. The section of stone described in the *Evening News* measured 70 feet in length, 22 inches in width, and 7 feet in thickness. Half of a keg of powder was used in this instance, although a more intense variety of explosive, "giant powder," was sometimes used in quarries of poor quality stone. In 1870, Harrison paid $5,000 in wages to the 12 men and two youths employed at his quarry. The Industrial Statistics of the Ninth U. S. Census for 1870 show that Harrison produced 6,600 feet of hammered stone valued at $4,000; his equipment a forge, a derrick, and the muscles of his laborers.

At the southern end of Eastern Avenue, David Beattie, the eldest son of William Beattie, in partnership with Arthur Wilcox, opened a quarry in 1893. Their operation was comprised of four separate pits in the area around the intersection of Eastern Avenue, McGowan, East Warren, and Kerr (Martine) Streets. Prior to opening his own quarry, David had been superintendent of his father's Bedford Street ledge.

David Beattie dropped out of high school after a year to work in his father's quarry. Education in the public schools diverted him from the apprenticeship in stone masonry, a path that had been followed by his father, his uncle and his cousin. He did, however, manage to acquire the skills of a quarryman, or general laborer. His younger brothers, William and Roy, attended Brown and M.I.T., respectively. The Beattie and Wilcox quarries contained the two basic varieties of Fall River granite, gray and pink. However, David Beattie's "pink granite" was darker in hue than that found in his father's quarry.

Beattie and Wilcox granite was used in the construction of the "Stone Bridge" to Aquidneck Island, the Fall River Public Library (1898-1899), and the state armory in Fall River (1895-1897).

Considering that many mill corporations quarried granite at the site of mill construction, Beattie and Wilcox supplied the stone for and built a surprising number of Fall River textile mills, including the Granite Mill No. 3 (1893), the Parker Mill (1895), and the Flint Mill No. 2 (1909). The firm also filled several outside contracts for granite construction: the weaving mill at the Bourne Mill complex in Tiverton; a war college on Coaster's Harbor Island, Newport, Rhode Island; a "Parker Mill" in Warren, Rhode Island; and a mill at South Manchester, Connecticut. Among Beattie and Wilcox's contracts specifying brick construction were the former Kerr Thread Mills No.1 and No. 2. The firm also built two noteworthy brick structures on Rock Street, the Second Bristol District County Courthouse (c.1905), and the former Truesdale Clinic (1913).

The Beattie and Wilcox Quarry was one of several commercial quarries, most of them located in the eastern part of the city, operating in Fall River in 1900. Arthur Wilcox died in 1914. David Beattie died three years later, at the age of 63. By 1923, the Beattie and Wilcox Quarry was idle. In 1928, the Careys reactivated the quarry site they had acquired from the David Beattie heirs.

The Beattie and Wilcox Quarry, like the Beattie and Cornell Quarry at the Narrows (owned by David's younger brother, William) and the Chauncey Sears Quarry, were owned

The Rolling Rock

fig. 12.4 **The Rolling Rock on the edge of the quarry in its original setting**

During the last Ice Age, about 100,000 years ago, glaciers about one mile in thickness broke away huge boulders from bedrock, rolled them to a smoother surface and, when the glaciers later retreated northward, deposited these boulders at new locations. These boulders are called "glacial erratics" or "perched" boulders. Glacial erratics range in size from small rocks to the size of Fall River's Rolling Rock, which weighs 140 tons. By the composition of the rock, some geologists estimate that the Rolling Rock originally came from a site in Dighton. This rock is a conglomerate boulder placed on granite bedrock.

Only a tiny fraction of these glacial erratics are "balanced rocks," that is, balanced on a fulcrum that allows them to be moved. Folklore has it that Native Americans used the Rolling Rock to extract information from captives by rolling the rock over their limbs. As early as the 1840s, geologist Edward Hitchcock feared that a nearby quarry might lead to the ultimate demise of the Rolling Rock. However, the quarry owners helped to prevent the rock from rolling into the quarry and, in 1930, the city preserved the rock after a 25-year struggle to decide its fate.

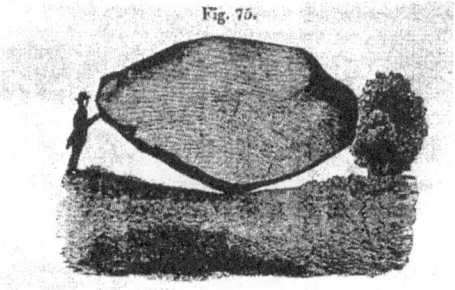

fig. 12.5 **Early engraving of the Rolling Rock**

and operated by building contractors who found it desirable to have a readily available supply of stone for their construction projects. Beattie's Ledge near Bedford Street, and the Harrison Quarry at Rolling Rock were examples of quarries operated, for the most part, to fill orders from outside customers.

As previously mentioned, many mill corporations operated their own quarries. The Merchants Mill No. 1 (1866), which once occupied the entire area between 13th, 14th, Bedford, and Pleasant Streets, was built with granite quarried on the site; additional stone was supplied by Beattie's Ledge. Two South End mill corporations, the King Philip Mills (1871-1892), and the Osborn Mills (1872-1886), maintained their own quarries and were constructed of granite. Another textile mill in the South End, the Montaup, was built of brick. Close by, another brick mill, the Slade, operated a quarry that produced foundation stone. The Slade Mill Quarry reportedly supplied granite for the construction of St. Patrick's Church (1881-1889). The Robeson Mill, located near the center of the city, also owned a quarry yet was built of brick.

In the 1920s there were more than 1,000 quarries operating in the United States, with more than a dozen localities where quarries, mills and shops were part of the local economy. Active quarries now number in the hundreds, with few of these integrated areas still in existence.

13 The Underground Railroad in Fall River

One of the most inspirational chapters in Fall River's history is that of the Underground Railroad in the city, most of whose activities took place in the vicinity of the Quequechan River and its falls. In *Bound for Canaan*, Fergus Bordewich observes that:

> The story of the Underground Railroad is an epic of high drama, moral courage, religious inspiration, and unexpected personal transformations played out by a cast of extraordinary personalities who often seem at the same time both startlingly modern and peculiarly archaic, combining then-radical ideas about race and political action with traditional notions of personal honor and sacred duty. ... It was the country's first integrated civil rights movement, in which whites and blacks worked together for six decades before the Civil War, taking great risks together, saving tens of thousands of lives together, and ultimately succeeding together in one of the most ambitious political undertakings in American history. Their collective experience is, if anything, an even greater record of personal bravery and self-sacrifice than is generally known. ... In an era when emancipation seemed subversive and outlandish to most Americans,

the men and women of the underground defied society's standards on a daily basis, inspired by a sense of spiritual imperative, moral conviction, and, especially on the part of African American activists, a fierce visceral passion for freedom. (11-4)

Prior to 1790, slaves cultivated tobacco on Virginia plantations and rice in the Carolinas. Cotton production, however, was hampered by the tedious process required to clean short staple cotton, whose seeds clung hard to the fiber. It took one worker a full day to clean a pound of cotton.

The invention of the cotton gin by Eli Whitney, a native of Westborough, Massachusetts, revolutionized the cultivation of cotton. Patented in 1794, Whitney's cotton gin, aided by power from a water wheel, could now do the work of hundreds. The large-scale cultivation of cotton was now economically possible.

With the development of the Arkwright spinning frame in England in 1769 and, later, the Cartwright power loom in 1785, the new English cotton textile manufacturing industry created a growing market for American cotton. Exports of cotton to England grew from virtually nothing in the early 1790s to six million pounds by 1796 and to 20 million pounds by 1801, increasing rapidly thereafter. (11-42)

With the development of textile manufacturing in New England after 1812, the demand for Southern cotton grew exponentially:

> The phenomenal expansion of the cotton economy carried slavery with it across the coastal states, through the still half-settled Mississippi Territory, and beyond, until the 'Cotton Kingdom' stretched from the Atlantic coast to Texas ... Mississippi production alone would swell from 20,000 bales in 1821, to 962,006 bales in 1859, almost one-quarter of the nation's output. And in each of the four decades before 1840, the slave population of Mississippi more than doubled. (11-42)

Textile manufacturing in New England was now inextricably bound to the Southern cotton plantation economy—and to slavery.

Contrary to what is commonly believed, the abolition of slavery was not popular in the North before the Civil War. While there was considerable sentiment that slavery was an evil institution, the prevailing opinion was that it would wither away over time. Any attempt at forcing the end of slavery, it was believed, would only antagonize the South and lead to civil war and the end of a united Republic. Few individuals believed that freeing the slaves was worth this risk.

In addition, as the cotton textile industry grew in New England, influential Northern textile mill owners became increasingly reluctant to threaten their supply of cotton by antagonizing Southern cotton plantation owners. War with the South would be ruinous for Northern mill owners and for everyone associated with the rapidly expanding textile industry.

fig. 13.1 **Arnold Buffum, anti-slavery leader, in 1826**

Arnold Buffum was devoted to the anti-slavery cause and traveled and lectured continuously on behalf of emancipation until his death. His mission took him throughout New England and the eastern seaboard and as far west as Ohio and Indiana. Between 1825 and 1831, he took several trips abroad to enlist the sympathies of the English and French. While in Paris, he became the friend of the Marquis de Lafayette and, in England, Ann, Duchess of Sutherland, who became very interested in his anti-slavery work.

Graphic source: Lowell: *Two Quaker Sisters*

This was also true of the Quakers, whose early prominence in Colonial and Federal Era shipping and related commercial activity led to their dominance of cotton textile manufacturing in Rhode Island and Fall River. Most Quakers were therefore extremely reluctant to tolerate the few abolitionists among them, not only because of the risk to their commercial interests but also because of the social insularity among the Friends that discouraged public activism. (54-116)

Nonetheless, a small minority of Quakers began to agitate for the end of slavery during the 1830s. One of the most prominent of these agitators was Arnold Buffum of Smithfield, Rhode Island. As a young boy, Arnold Buffum had learned to despise slavery by the stories told him by his family's old escaped slave, who related the horrors of being captured in Africa and being brought to America on a slave ship. Arnold later became a farmer, a manufacturer of hats, an inventor and a lecturer.

In 1824, Arnold Buffum moved his wife and five daughters to Fall River. His daughters were Sarah, Elizabeth, Lucy, Rebecca, and Lydia. During his years in Fall River, he became increasingly involved with the abolitionist movement and became one of its leaders. He traveled throughout New England speaking against slavery and, with William Lloyd Garrison and others, founded the New England Anti-Slavery Society in Boston in 1832 and became its first president. (54-xxii)

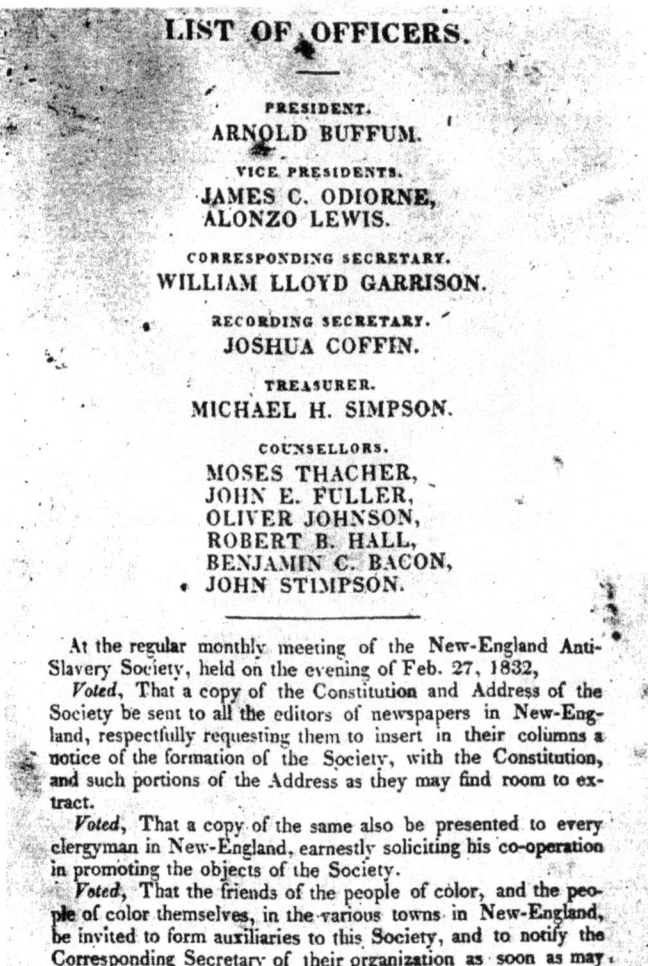

fig. 13.2 **Officers of the New England Anti-Slavery Society**

This reproduction of the list of officers of the New England Anti-Slavery Society in February, 1832 shows Arnold Buffum as the first president of the organization.

Graphic source: Lowell: *Two Quaker Sisters*

His daughter Sarah married Nathaniel B. Borden, manufacturer and third mayor of Fall River. Another daughter, Elizabeth, married Samuel B. Chace, also of Fall River, and subsequently moved to Valley Falls (now Central Falls), where her husband founded a major cotton manufacturing enterprise.

Arnold Buffum inspired his daughters with his anti-slavery zeal, and they became converts. However, as Elizabeth would later say, "we found very few people who were ready to give any countenance or support to the new anti-slavery movement ... their commercial relations, their political associations, and with many, their religious fellowship with the people of the South, so blinded their eyes, hardened the hearts and stifled the consciences of the North." Even Quakers "had become so demoralized, that they too, with rare exceptions, shut their eyes to the great iniquity." Because of their anti-slavery activity, Arnold Buffum, his daughters, and their families were disowned by the Society of Friends. Except for the few abolitionists, the general feeling among Quakers was that "on the question of slavery, silence should be maintained." (54-123)

In 1834, the Fall River Anti-Slavery Society was formed and, in 1835, Elizabeth and Sarah Buffum and 25 other women organized the Female Anti-Slavery Society in Fall River. The Buffums were also instrumental in organizing an anti-slavery convention in Fall River.

As the number of escaped slaves increased after 1830, the Underground Railroad was created to safely move them from the South through the North and on to Canada. The homes of the Buffum sisters, Sarah in Fall River and Elizabeth in Valley Falls, became stations on the "railroad." As Elizabeth related later in 1891, in *Two Quaker Sisters*:

> From the time of the arrival of James Curry at Fall River, and his departure for Canada, in 1839, that town became an important station on the so-called underground railroad. Slaves in Virginia would secure passage, either secretly or with consent of the captains, in small trading vessels, at Norfolk or Portsmouth, and thus be brought into some port in New England, where their fate depended on the circumstances into which they happened to fall. A few, landing at some towns on Cape Cod, would reach New Bedford, and thence be sent by an abolitionist there to Fall River, to be sheltered by Nathaniel B. Borden and his wife, who was my sister Sarah, and sent by them to my home at Valley Falls, in the darkness of night, and in a closed carriage, with Robert Adams, a most faithful friend, as their conductor.
>
> Here we received them, and, after preparing them for the journey, my husband would accompany them a short distance, on the Providence and Worcester Railroad, acquaint the conductor with the facts, enlist his interest in their behalf, and then leave them in his care. They were then transferred at Worcester to the Vermont road, from which, by a previous general arrangement, they were received by a Unitarian clergyman named Young, and sent by him to Canada, where they uniformly arrived safely. I used to give them an envelope, directed to us, to be mailed in Toronto, which, when it reached us, was sufficient by its

Alfred J. Lima

THE ANTI-SLAVERY RECORD.

VOL. I.   SEPTEMBER, 1835. [Second Edition.]   NO. 9.

THE DESPERATION OF A MOTHER.

"Why do you narrate the extraordinary cases of cruelty? These stories will not convert the cruel, and they wound the feelings of masters who are not so."
REPLY. Cruelty is the fruit of the system.

In Marion Co., Missouri, a Negro-Trader was, not long ago, making up a drove for the Red River country. He purchased two little boys of a planter. They were to be taken away the next day. How did the mother of the children feel! To prevent her interference, she was chained in an out-house. In the night she contrived to get loose, took an axe, proceeded to the place where her [yes, *her*] boys slept, and severed their heads from their bodies! She then put an end to her own existence.

☞ The negro-trader and planter quarreled, and went to law, about the *price!*

fig. 13.3 **An incident described in the *Anti-Slavery Record***

Stories such as this inflamed the passions of the abolitionists.

Graphic source: Lowell: *Two Quaker Sisters*

postmark to announce their safe arrival, beyond the baleful influence of the Stars and Stripes, and the anti-protection of the Fugitive Slave Law. (54-127 and 128)

The passage of the Fugitive Slave Law in 1850 made it illegal to harbor slaves in the North, and allowed their Southern owners to retrieve escaped slaves wherever they were. Anyone harboring fugitive slaves was now subject to heavy fines and imprisonment. The Bordens in Fall River and the Chaces in Valley Falls were determined to disobey the law, whatever the consequences, and "our children and servants entered heartily into our sentiment, although some of our Christian neighbors did not."

fig. 13.4 **Samuel G. Chace and Elizabeth Buffum Chace**

Both Mr. and Mrs. Chace were residents of Fall River who, following their marriage, moved to Valley Falls (later Central Falls), Rhode Island, where Samuel Chace established a successful cotton textile manufacturing enterprise. Samuel and Elizabeth operated a way station on the Underground Railroad from their home for many years, in coordination with another way station in Fall River operated by her sister Sarah Buffum Borden and her husband Nathaniel B. Borden. In 1891, Mrs. Chace wrote her memories of that time in "My Anti-Slavery Reminiscences," which her nephew Malcolm Lovell later incorporated into a book published in 1937 titled *Two Quaker Sisters.*

The fact that the Commonwealth of Massachusetts would now prostrate itself to the Southern slave owner and be complicit in returning free men into slavery infuriated abolitionists such as Henry David Thoreau. Following the capture and return of a fugitive slave by the Boston and Massachusetts authorities, Thoreau wrote in *Slavery in Massachusetts*:

> I did not know at first what ailed me. At last it occurred to me that what I had lost was a country. I had never respected the Government near to which I had lived, but I had foolishly thought that I might manage to live here, minding my private affairs, and forget it. For my part, my old and worthiest pursuits have lost I cannot say how much of their attraction, and I feel that my investment in life here is worth many percent less since Massachusetts deliberately sent back an innocent man, Anthony Burns, to slavery. I dwelt before, perhaps, in the illusion that my life passed somewhere only between heaven and hell, but now I cannot persuade myself that I do not dwell wholly within hell. (78-29)

Elizabeth Buffum Chace relates one instance where a slave had escaped from Virginia with his wife and child and found employment in New Bedford. His former owner tracked him down to New Bedford, "but the colored people of that town discovering the purpose of the searchers, communicated with some of the few Abolitionists, and the man was hurried off to Fall River, before the man-stealer had time to find him; and my sister Sarah and her husband Nathaniel Borden, dressed him in Quaker bonnet and shawl, and sent him off in the daylight, not daring to keep him till night, lest his master should follow immediately." (54-129)

Another instance involved a former slave who fled from Maryland with her family, established herself as a laundress and made a secure home in Fall River. However, a known constable was seen prowling the neighborhood in the city where colored people lived and suspiciously peering into the stable where her older son once worked. The abolitionists in Fall River hurried the woman

fig. 13.5 **Nathaniel Briggs Borden**

Source: Courtesy of the collection of the Fall River Historical Society

and her three children off in a carriage driven by Robert Adams to the home of Samuel and Elizabeth Chace in Valley Falls. There the fugitive family awaited the delivery of household effects and the arrival of the eldest son, who was then working on a farm. "On the third or fourth day, the boy arrived with money from the good friends at Fall River, and we sent them off, still fearing their capture on the road." However, the envelope finally arrived at the Chace home, postmarked from Toronto, signifying that they had arrived safely "and the man-stealers lost their prey." (54-133)

In his *History of Fall River*, Phillips mentions that a very prominent Quaker, Israel Buffinton, stabled three horses, two of which were kept in continuous use between Fall River and the next depot. (65-133) There were many other families that harbored fugitive slaves in Fall River, but history has not recorded their activities.

Slaves escaped by land principally from the border states. It was almost impossible for a slave to escape from the deep South by land, due to the extent of hostile territory to be traversed for months before reaching safety. Escape from the deep South was therefore easier and quicker by sea, where a Northern port could be reached in a week. Northern ships were considered a commercial extension of the North, and stowaways—with or without the consent of the captain—frequently found a safe berth on a northward bound vessel. The consequences of being caught harboring a fugitive slave, however, could be very severe for the captain. This escape by sea became known as the "Saltwater Underground." (11-271)

One of the southern terminals for sending slaves to New Bedford and then to Fall River was Portsmouth, Virginia. In that city, a colored woman named Eliza Baines worked for sea captains and therefore knew the destinations of vessels and their times of sailing. She hid escaping slaves in her home and, when conditions were right, sent them off on ships bound for New Bedford or Boston. (71-7)

After the firing on Fort Sumter and the start of the Civil War, abolitionists were blamed for antagonizing the South, and mobs in Northern cities threatened and attacked the abolitionists, as Elizabeth Buffum Chace describes occurring in Valley Falls in *Two Quaker Sisters*. In his 1877 *Centennial History of Fall River*, Henry H. Earl, in his description of Nathaniel Briggs Borden as the third mayor of Fall River, indicates that the anti-abolitionist mobs also existed in Fall River:

> He was among the early and prominent friends of the slave, and assisted many a fugitive, either directly or indirectly, on his road to freedom. At a time when it was fashionable to mob Abolitionists he opened the Washington School-House, then his private property, in which to form an anti-slavery society. (25-229)

The Underground Railroad in Fall River must have been quite extensive, since escaped slaves arriving by sea at Cape Cod ports, Wareham, and at New Bedford were funneled through Fall River on their way to Canada. Some slaves arrived in Fall River directly by ship. In his "The Underground Railroad in Massachusetts," William H. Siebert says that,

Alfred J. Lima

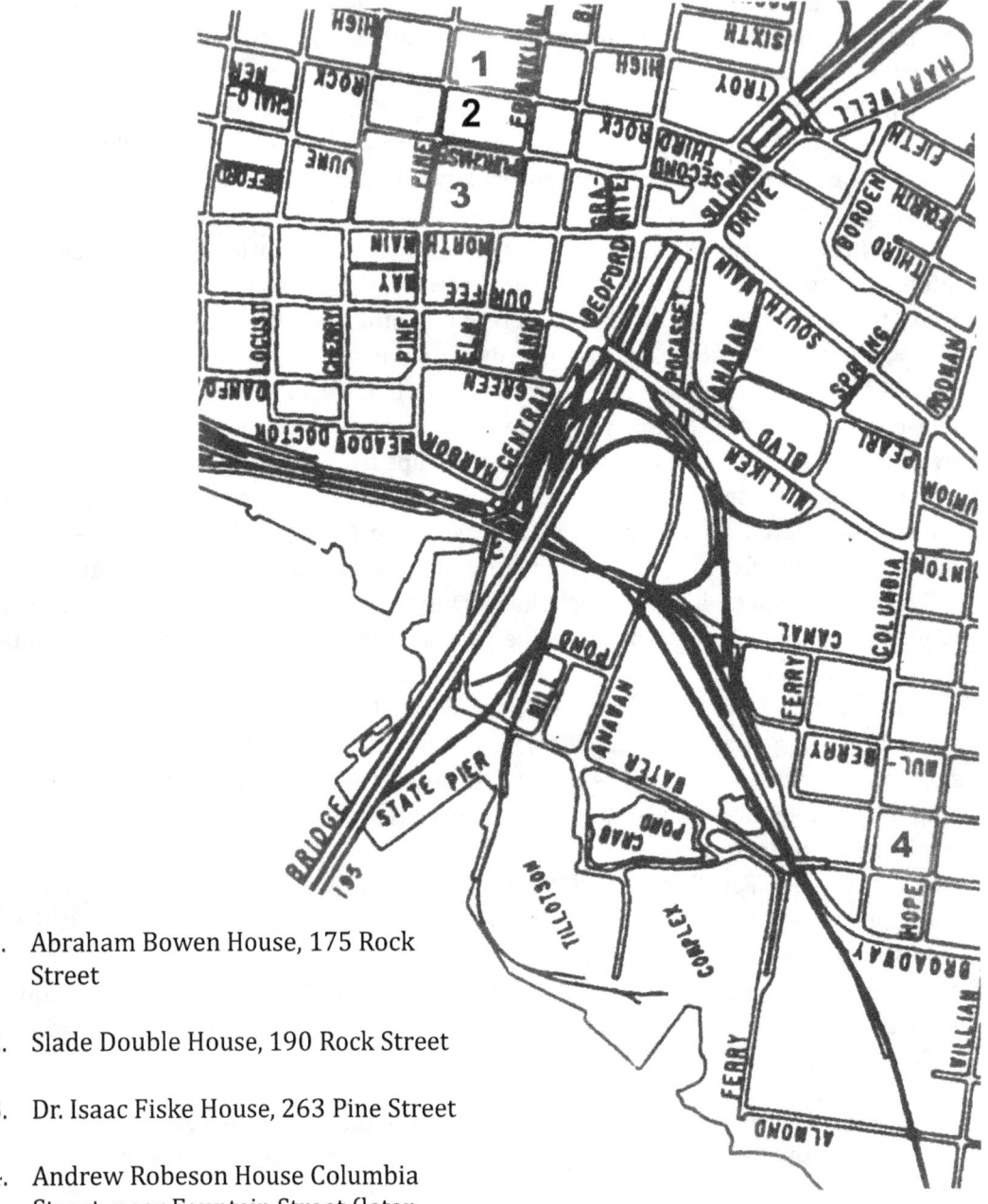

1. Abraham Bowen House, 175 Rock Street

2. Slade Double House, 190 Rock Street

3. Dr. Isaac Fiske House, 263 Pine Street

4. Andrew Robeson House Columbia Street, near Fountain Street (later moved to 451 Rock Street)

5. William B. Canedy House, 2634 North Main Street (not shown on the map), opposite Canedy Street and the Wiley Elementary School

fig. 13.6 **Known Underground Railroad way stations in Fall River**

A RIVER AND ITS CITY

from Fall River's underground way stations, there were four routes that escaped slaves took to Canada. Two went northwesterly, one to Valley Falls and one to Pawtucket, from which slaves were placed on the Providence and Worcester Railroad to Worcester and on to Vermont. Of the northerly routes, one went by land to Norton and on to North Attleboro and Medfield, and the other proceeded along the Taunton River to Taunton and then via Framingham to Concord. (71-10) The Underground Railroad in Concord included luminaries such as Bronson Alcott and Henry David Thoreau.

The best-known way station in Fall River is the building that is the Fall River Historical Society, now located at 451 Rock Street but originally built near lower Columbia Street, near Fountain Street, once a fashionable residential area. It was built in 1843 by Quaker Andrew Robeson for his son. Its movable bookcase, which once led to a secret underground hiding place, is still visible and operable at the Historical Society.

Three of the other four known way stations in the city are all within a short distance from the Quequechan River in the heart of the city: the Abraham Bowen House at 175 Rock Street, the Slade Double House at 190 Rock Street, on the corner of Purchase Street, and the Dr. Isaac Fiske House at 263 Pine Street. Dr. Fisk was known to have hidden runaway slaves in his attic. The fourth site is further north near Steep Brook at 2634 North Main Street, the William B. Canedy House.

Having done their part to free the slaves, the Buffum sisters in Fall River and Valley Falls

fig. 13.7 **A flag made by Arnold Buffum Chace**

—grandson of Arnold Buffum and son of Samuel and Elizabeth Buffum Chace—as a boy of 11 in 1856 during the Fremont presidential campaign. Arnold Buffum Chace was later Chancellor of Brown University from 1907 to 1932.

Source: Lovell: *Two Quaker Sisters*

then turned their attention to the rights of women and became leaders in the Women's Suffrage movement. In 1870, Elizabeth Buffum Chace co-founded the Rhode Island Women's Suffrage Association and served as its president for almost 30 years until her death in 1899. She ends "My Anti-Slavery Reminiscences" in *Two Quaker Sisters* in quiet defiance and a call to duty (54-182):

> All of these experiences were an important feature in the education of our children of which, circumstances being as they were, I would by no means have deprived them. For there is no better influence toward the building up of a strong virtuous manhood and womanhood, than the espousal in early life of some great humanitarian cause as a foundation. By such preparation men and women are made ready to take up all questions which concern the advancement of mankind. The slavery of the black man is abolished, the shackles have fallen from his limbs and he is crowned with the diadem of citizenship. It is too late to become an Abolitionist now but in the process of overthrowing one great wrong there is always laid bare some other wrong, which requires for its removal the same self-sacrificing spirit, the same consecration to duty, as accomplished the preceding reform. So it has ever been.
>
> Valley Falls, Rhode Island, 3rd month, 3rd, 1891.

fig. 13.8 **Routes of the Underground Railroad in Eastern Massachusetts and Rhode Island**

Fugitive slaves were brought from their landing stations on Cape Cod, Wareham, and New Bedford to Fall River. From Fall River, they were sent to three principal interim destinations: Valley Falls, Pawtucket, and Taunton.

Reproduced from Siebert: "The Underground Railroad in Massachusetts"

14 The People of the Quequechan River Valley

Working Conditions in the Early Mills

Following the end of King Philip's War in 1676, a new permanent settlement began to grow along the falls of the Quequechan River. Colonel Benjamin Church, victor over King Philip, built a sawmill and a grist mill on the Quequechan in 1680 with his brother Caleb, a millwright from Watertown. The farming community that grew below the hill—on the flat land between the Taunton River and the base of the hill (now known as the North End)—changed little during the 130 years from 1680 to 1810.

With the development of the early water-powered textile mills after 1813, the area around the falls began to change. The operatives in the early mills were mostly Yankee women and children from local farms. As the mills increased in number, size, and capacity, the local supply of surplus farm labor was inadequate to meet the demand.

In 1820, there existed in the village the two original mills on the Quequechan River, the Fall River Manufactory and the Troy Cotton and Woolen Manufactory. The population of the village was 1,594 residents and the mills employed 850, counting the numerous home-based outwork weavers that took the yarn from the mills and wove it into cloth on their cottage looms.

Within seven years, however, the village had become transformed. There now existed on the Quequechan River six cotton factories, two cloth printing mills and the Fall River Iron Works. Population since 1820 had increased by 500 persons and inside the mills themselves approximately 700 persons were employed. Most of the workers in the cotton mills were women and children under 16. As the demand for specialized skills such as block printing increased, the immigration of skilled workers from England and Scotland began. Of the total work force of 700 in Fall River in 1827, only 42 were English, Scottish, or Irish immigrants, and most of these were employed as block printers. (10-48)

In the 1840s, a combination of the Potato Famine in Ireland and the need for new hands to work in the mills and to build the railroads resulted in the immigration of the Irish into the United States. This Irish immigration would continue until after the turn of the century.

The typical workday in Fall River's textile mills for many years was 14 hours, from 5:00 in the morning to 7:30 at night, for men, women, and children. The work week was 84 hours, six days a week, with only the Sabbath off. The 14-hour work day remained in effect for approximately 60 years, until the Massachusetts legislature finally passed a 10-hour law in 1875. It would be more years, however, before this law was strictly enforced.

In the early Fall River mills, as elsewhere where the Rhode Island model prevailed, child labor, under close supervision, was used in spinning mills to tend the throsle frames and then deliver the yarn to families in the countryside who would weave it into cloth. (10-24) Later, they worked in the integrated mills. Children became mill workers as soon as they developed stamina, or about eight years old.

One of these mill operatives, Benjamin Pearce, was required to go to work at the age of eight because of family financial needs. He worked 14 hours a day, 84 hours a week, for 50 cents a week, or less than a penny an hour. In his later years, Pearce (through his daughter) recounts his first day at work at a Fall River textile mill:

> Bright and early on a Monday morning in March, 1827, Benjamin, with his clean apron on and a head erect, trotted off to the factory. When, an hour and a half later, he went home to his breakfast, he was enthusiastic over his new employment; at noon it was noticed he was a little less so. It was a woolen factory for the manufacture of sattinet, the first remove from the old-time method of carding, spinning and weaving wool by hand. His employment was splicing the rolls, and kept him upon his feet and walking to and fro through the fourteen hours of the day. Hence it was not strange that as the day began to wane, and his knuckles to bleed from the constant abrasion in splicing the rolls on a rough apron, himself ready to drop upon the floor from weariness, that his spirits began to flag and his work to get away from him. For this the only evidence of pity on the part of the man in charge of the work, was a stunning blow with the flat of his hand on the ear, with an admonition to "wake up."

> When, at the expiration of fourteen hours, Benjamin went out of the factory, it was with less exuberance of spirits than when he entered it in the morning. All the knuckles of his right hand were bleeding, his face and flaxen hair were begrimed with the dye-stuff and oil from the wool he had been handling, the spirit had all gone out of him, and he was anything other than the bright, cleanly boy who went forth with such towering ambition in the morning. He related, amid bitter sobs, the brutal treatment he had received at the hands of Durfee, the young understrapper under whose charge he had been placed. His mother consoled him as only a mother can, and he sat down to his well-earned supper. He had not, however, finished the meal, when he leaned forward sobbing upon the table and in a moment was asleep, sleeping soundly until the next morning, his mother having removed his clothing and put him to bed. (62-35)

Pearce notes in his reflections that "probably plantation slaves of similar age were treated with more humanity, or with less cruelty, at least, than were many of the boys in this instance."

Occasionally, the brutality brought on resistance, as when a boy locked himself in a closet to escape the wrath of the pursuing overseer. He escaped through a floor board by grasping at timbers above the rushing Quequechan River and made his way home and hid under his bed. When the closet door was opened, the overseer panicked, fearing that the boy had chosen to fall into the river and drown rather than face the punishment that was in store for him. An alarm was given, the factories stopped and the Quequechan drawn down to find the boy.

> While this was progressing, tidings of the event had been conveyed to his home, and the lamentations of the family, wafted to him in his hiding place, were the first intimation he had of what a commotion he had created. When he crept down and showed himself, the family could scarcely credit their senses. Word was at once passed along the line and the searchers recalled. It is needless to add that the little fellow enjoyed his well earned exemption from punishment, and the mill hands their two hours' recess. (62-40)

Pearce relates another instance where, threatened with punishment for a minor infraction by an overseer, a young worker fitted out a leather patch with tacks protruding outward, which he attached under his clothing. When the overseer struck, and struck again, he quickly removed his bleeding hand.

> Quickly relinquishing his hold on the boy, he ordered him to put on his coat and leave the mill, at the same time wiping the blood from his hand. The lad did as he was bidden, but he had the satisfaction of seeing the superintendent on the street the next four or five days with a white cloth around his hand. (62-41)

In *Constant Turmoil: The Politics of Industrial Life in Nineteenth-Century New England*, Mary H. Blewett describes how this treatment extended even to the mill owners' own children:

> Mill owner David Anthony treated his own sons, Jim and Fred, much the same way. During the long week he worked them hard without any pay, along with the rest of the 'help' of operatives, then insisted on the strictest observance of the Sabbath. His sons were confined to the church or house all day. When the more defiant son asked his father for a day off, the answer was no. Fred then 'went down into the wheel pit [of the mill] and cast a large spike into the gearing ...' Repairs took several days. Fred got a beating as well as a vacation. Both boys frequently ran away from their father's mill to Boston, and Jim once shipped out aboard a merchantman to China, returning after a few months. Anthony's training of his sons gave them no quarter of any kind. (10-49)

Textile mills operated as a tightly-orchestrated system, and therefore punctuality was essential. The discipline of time was enforced by the mill bell, at a time when clocks were a rarity in workers' homes and alarm clocks were unknown. The first bell rang at 4:40 am. At the ringing of the 5:00 am bell, operatives were to be at their places and ready to begin work. Being late even a minute meant serious consequences. At 5:10, the gates to the mill were shut.

All etchings and early photographs of mill buildings show a high wrought iron fence surrounding the mills, erected to keep workers on the property and to control the time of arrival of workers and the children who brought them breakfast and "dinner" at 12:30.

The mill buildings themselves were often locked following the morning bell, to keep operatives in the building. This common practice became an issue when a fire in Fall River's Granite Mill in 1874 resulted in 40 deaths and 80 serious injuries.

Breakfast bells rang at 7:00 am, when the gates were open to allow children to bring their parents or siblings a breakfast. At 7:30, the bell rang again to start work. At some early Fall River mills, at 11:00 am, the male operatives were treated to New England rum. (25-28)

Dinner bells rang at 12:30 pm, when the gates were again open to allow children to bring their parents or siblings their "dinner." Gates were again closed at 1:00 pm.

At 7:30 pm, the evening bell rang, signaling the end of the work day and the opening of the gates.

The early romance of the model communities in the Waltham-style mills soon gave way to disillusionment. In 1841, a young girl expressed her frustration with the regimentation of mill life in Lowell:

> I shall go home, where I shall not be obliged to rise so early in the morning, nor be dragged about by the factory bell, nor confined in a close noisy room from morning to night. I shall not stay here ... Up before day, at the clang of the bell,—and out of the mill by the clang of the bell—into the mill, and at work in

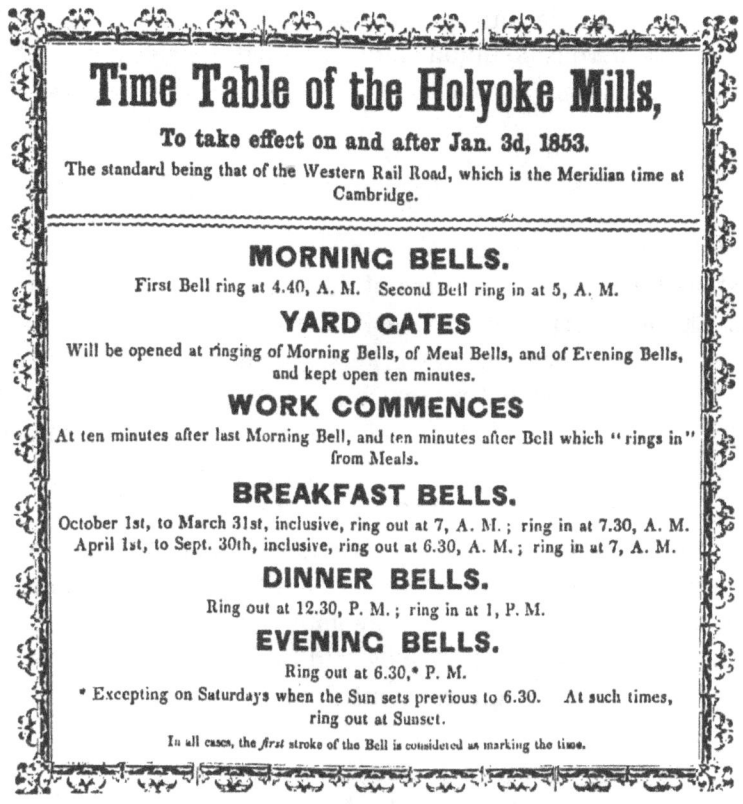

fig. 14.1 **Bell schedule for a New England textile mill, 1853**

Source: Dunwell: *Run of the Mill*

obedience to that ding-dong of a bell—just as though we were so many living machines. (24-49)

The regimen imposed by the overseers and the harsh treatment of mill operatives by their supervisors was a source of constant discontent. In *Constant Turmoil*, Mary H. Blewett describes these conditions in Fall River:

> Most mill women endured harsh language, insults, sexual harassment, long hours, and low wages. They often complained that they had to eat their meals in such a hurry that, in the words of one, 'It hurts me.' Some were in such fear of missing the factory bell that they could not sleep and rose in the middle of the night. Children cried out in their sleep in fear of missing the 5:00 AM deadline. One female operative complained that the overseer's language was 'exceedingly profane ... It makes me tremble ... We live all the time in fear of their abuse.

Another said, 'I bear it [the abuse] as well as I can ... I know we ought to ... [rise up against it], but there is no union among us.' High turnover rates or poverty made organized resistance difficult. (10-49)

Housing of Mill Operatives

The Rhode Island system of cotton textile manufacturing developed by Samuel Slater along the Blackstone River Valley consisted of the development of small mills on modest water privileges. These waterfalls were usually located in remote rural areas, requiring mill developers to build housing for their workers. The dwellings were often basic wooden tenements that became overcrowded but which were often better than poor families could find elsewhere. The tenements were clustered adjacent to the mill, so that everyone could walk to work and live within earshot of the mill bell. Residence in mill housing was usually mandatory, but few objected and alternative housing was scarce, if available at all.

In Fall River during the early years, mill housing was not always necessary because the Yankee workers came from the village or local farms, although the first mills (the Fall River Manufactory and the Troy) did construct a few four-unit dwellings nearby. However, the erection of new and larger mills in the 1820s, and the immigration of foreign operatives beginning in the 1840s, required the construction of mill housing on either side of the

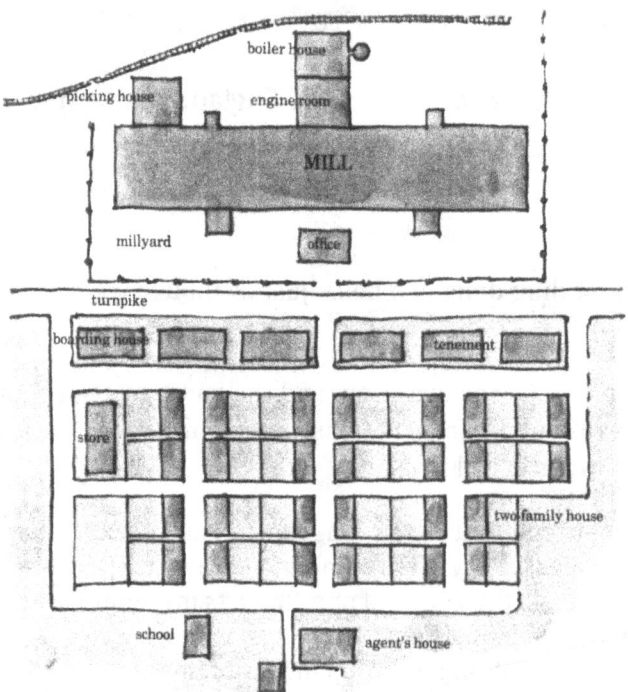

fig. 14.2 **Typical layout of a Rhode Island system mill village**

Source: Macaulay: *Mill*

Quequechan River adjacent to the mills. Often the mill owners would retain ownership of the mill housing, deducting rent from operatives' pay. Later, the mills sold the tenements to others to rent them out.

This early mill housing (in common with all housing in Fall River) had no plumbing, since there was no city water supply until 1875 to supply domestic water to dwellings. Water for household use was either obtained from the Quequechan River (it was sold by the pail at "The Landing" at Hartwell Street) or from shallow wells. Every back yard had its outhouse for the use of residents of the crowded tenements. These tenements had no electricity, since electricity did not come into general use in the city's neighborhoods until after 1920.

The layout of the Richard Borden Mill on Rodman Street at the corner of Plymouth Avenue was a classic example of mill and housing juxtaposition. On the opposite side of Rodman Street, row on row of identical tenements were constructed in a grid arrangement. All of this substandard housing—including all of the housing in the area of what is now Milliken Boulevard below South Main Street—was demolished later as part of a public urban renewal effort. A classic layout of the Rhode Island mill village is shown in figure 14.2.

Around 1877, mill developers in Fall River ceased building housing themselves and began contracting housing construction to others. This relieved the mill owners of the expense and obligation of providing housing for their workers and resulted in better housing and more freedom of housing choice for new workers. (25-133)

Wages and the Company Store

In the Rhode Island System mills, including Fall River, wages per week in 1830 averaged $4.50 for males, $2.00 for females, and $0.67 for children. Payment was not in cash but in credit redeemable at the company store.

As described by Steve Dunwell in *Run of the Mill*:

> Mill stores served company interests by minimizing cash flow and making the indebted workers more dependent on the factory. Wide abuse of the practice justified its infamy. Work contracts required employees to make all purchases at the mill store. One Rhode Island mill posted a typical warning:
>
> *Notice: Those employed at these mills and works will take notice, that a store is kept for their accommodation, where they can purchase the best goods at fair prices and it is expected that all will draw their goods from said store. Those who do not are informed that there are plenty of others who would be glad to take their place at less wages.* (24-71)

Hannah Borden was one of the few workers who successfully resisted the company store. In 1817, at age 14, Hannah pleaded with her father—an investor in David Anthony's mill on the Quequechan River, the Fall River Manufactory—to allow her to work in the mill operating one of the three new and very novel power looms. Previous to that, Hannah had taken the yarn from the local spinning mills and wove it into cloth for home use and for sale by the mills. She worked a 14-hour day, from 5:00 in the morning until 7:30 in the evening, with a break for breakfast and lunch for six days a week. She was paid not in cash but in credit at the company store.

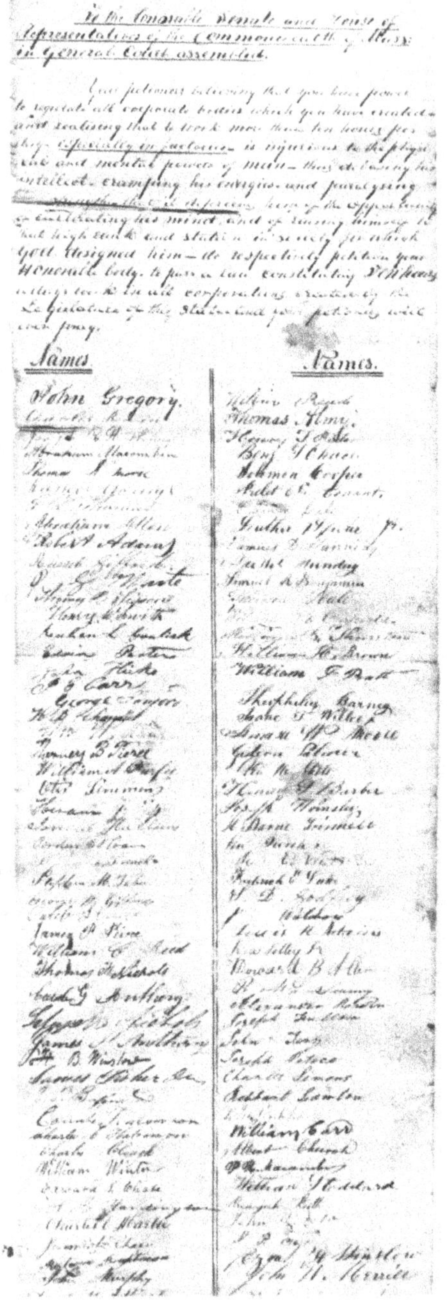

One day, Hannah found herself in debt to the store. She asked to see the books, but the clerk refused. Undaunted, Hannah appealed to Anthony, who allowed her to see her account, which showed charges for suspenders and rum. After another year of being dissatisfied with the quality of goods at the company store, Hannah asked to be paid in cash. Anthony at first refused, citing the precedent of paying an operative in cash and having to do the same for other workers. However, Hannah insisted and, given her father's role as an investor in the mill, she was finally—and quietly—paid in cash. (10-46)

Paying workers only in goods from the company store rather than in cash—and the abuses that arose from that system—was one of the causes of labor unrest in the early cotton mills.

fig. 14.3 Petition of Fall River mill workers for a 10-hour work day

Shown is the first two feet of a 10-foot petition presented to the Massachusetts Legislature in 1845.

Source: Dunwell: *Run of the Mill*

Labor Unrest and Improvement in Working Conditions

Early capitalism in the textile industry was one of boom and bust. Depressions occurred in regular 10-year cycles: in 1816, 1827, 1837, 1848, and 1857. Fortunes could be made, but they could just as easily be lost. Increased competition led to reduced profit margins. In 1834, when wages were cut by 15 percent, walkouts occurred in Lowell and elsewhere.

While unsuccessful in their efforts to restore their wages, it was the beginning of a long history of worker-management conflict in the New England textile industry.

The harsh working conditions of the mills eventually led to labor unrest and organized

fig. 14.4 **The "bread strike" of 1875**

Precipitated by two wage cuts in 1873 and late 1874, the strike resulted in the governor calling out the militia and allowing mill owners to fire strikers and reopen the mills on their own terms. The sign says "15,000 white slaves for sale at auction." Note the loaves of bread being carried on poles. The scene shown in the etching occurred above the Quequechan River, where it flows under the center of downtown.

Reproduced from: Blewett: *Constant Turmoil*

strikes. As early as 1832, mule spinners in the Anawan Mill went on strike. They tried to prevent management from replacing them, but they were unsuccessful and left the village as the strike collapsed. In 1840, block printers who struck the Robeson mill over the introduction of engraving machinery were fired and replaced. (10-49)

As advances in technology allowed speedups in work and additional workloads, the 14-hour day became increasingly intolerable. In 1842, Fall River and other mill town workers began to agitate for a 10-hour day. In that year, Fall River mechanics demanded a 10-hour workday in a petition to the Massachusetts legislature. Two years later, the Fall River mechanics helped to organize a region-wide convention to press for the shorter workday. That gathering, held in Boston in 1844, resulted in the formation of the New England Workingman's Association. (48-22)

Appeals to the Massachusetts legislature, heavily represented by industry interests, resulted in no action. The depression of 1848 left workers without employment and therefore little bargaining power. However, with the return to prosperity in 1852, workers began to gain political power and therefore effectiveness. Laws were passed in Massachusetts and Rhode Island regulating work hours, but legislators opposed to the measures introduced amendments that effectively made the shorter hours optional. Even child labor laws were so cleverly written as to be practically ineffective.

In 1848, English workers in the city were joined by native-born Yankees in opposing a surprise wage cut. Three of the leaders of the strike were indicted by a Fall River grand jury for creating "riot" and "terror" and were threatened with being blacklisted throughout New England. Blacklisting was routinely carried out against labor activists in the New England textile industry for many years. Two of the leaders were found guilty and jailed. Those who testified for the defense left the city, fearing the blacklist. Once again, a strike had failed. (10-85)

Another longer and even more acrimonious strike followed in 1850-51, but that, too, failed. In 1866, mule spinners led a work stoppage, but this time with weavers, carders, and all other textile workers, but to no effect. In 1867, workers once more struck for the 10-hour day, but again unsuccessfully. Spinners struck in 1875 in an effort to restore a wage reduction, but it "failed in the face of hunger, threatened evictions from company housing, and the presence of two companies of the Massachusetts militia protecting strikebreakers." (33-175)

During the 1873 depression, wages were cut twice and hours of work cut. In 1875, following another wage cut of 10 percent, a Fall River strike, led by the mule spinners and the weavers, was at first successful. However, after a bitter struggle, the strike ended when the governor ordered the militia to enter the city and allowed the mill owners to fire the strikers and reopen the mills on their own terms. These events incited labor protest throughout New England.

In *Labor Conflict in the United States*, Ronald L. Filippelli describes the situation in Fall River:

Like many similar actions of the period, the victory of the textile workers in 1875 was short-lived. Indeed, the owners had learned a lesson from their defeat. They, too, realized that solidarity was necessary for victory and formed the Board of Trade to coordinate a wage reduction and a city-wide lockout which began in September of 1875. After four weeks, the mills reopened to non-union workers who signed "yellow-dog" contracts agreeing not to join a union as a condition of employment. In the disaster of 1875, the weavers, carders, and loomfixers unions were destroyed. Only the spinners survived. (33-175)

However, this harsh treatment left a lingering resentment among the city's textile workers. As Tom Juravich described in *Commonwealth of Toil*:

In order to regain their jobs, the city's proud British workers were forced to sign 'yellow dog contracts' pledging to quit their unions. But they did not consider these contracts binding 'as it was their starving children that compelled them to sign ... against their wills.' 'An empty stomach can make no contracts,' George McNeill wrote. (48-45)

While English mule spinners and weavers were sought by the mills to work the new English textile machines, they came at a price for the mill owners in greater labor unrest. In his 1877 *Centennial History of Fall River, Massachusetts*, historian Henry H. Earl lets his frustration show at the situation:

fig. 14.5 **Young operatives in the Cornell Mill, Fall River, 1912**

Source: Photograph by Louis Hine in the Library of Congress.

> The operatives employed in Fall River are mostly foreigners, but the American, French and Irish elements are well disposed as a rule, and give little trouble except when led by the English (Lancashire) operatives, who, having come from the most discontented districts of England, have brought their peculiar ideas and the machinery of their home style of agitation along with them. This system is not relished by the other operatives, but so potent has been the influence of the active element that it has sometimes held the others in awe, and in times gone by has even been so powerful that if one of the trades-union men went into a mill and held up his hand, all the operatives at once, quitting their machines, left the mill, and went outside to find out why it was that they left their work. But it is hoped that the day of this style of terrorism and despotism has gone by, and that the compulsory system of school education, now in force in Massachusetts for factory children, will put them in a position to control their own motions, rights, and interests. (25-111)

But the strikes continued, one in 1879 and three between 1884 and 1894. After coming to Fall River in 1874, Lancashire agitator and weaver George Gunton observed that Fall River was "the hardest place for work and the meanest place for wages" in New England (5-131). Indeed, in the 18 years from 1884 to 1902, monthly wages actually *decreased* from $18.50 to $17.32. (20-253)

The calculating attitude of mill owners to their workers is reflected in remarks attributed to a prominent Fall River mill owner in 1855:

> As for myself, I regard my work-people just as I regard my machinery. So long as they can do my work for what I choose to pay them, I keep them, getting out of them all that I can. What they do or how they fare outside my walls I don't know, nor do I consider it my business to know. They must look out for themselves as I do for myself. When my machines get old and useless, I reject them and get new, and these people are part of my machinery. (10-39)

As Mary H. Blewett mentions in *Constant Turmoil*, the harsh working conditions of textile operatives were not particular to Fall River. In 1884, a Fall River weaver, who declared that he would rather spend five years in prison than remain working in the city, began tramping throughout New England searching for something better. In Connecticut, he found mills that were swindling their employees in company stores and where operatives worked from 5:00 am to 7:00 pm under conditions that were "nothing but slavery and oppression." After working for a while in Rhode Island mills, he returned to Massachusetts, convinced that there was no better alternative. (10-225)

Even the sainted Quaker abolitionist from Fall River, Samuel Chace, who moved to Valley Falls with his wife Elizabeth Buffum and established a station on the Underground Railroad, operated the Chace Mills in that community in a manner that was no different than other mills in the region. The Valley Falls mills, in the words of one of its operatives, "constituted a complete system of tyranny" with a spy network that prevented anyone from reading the union press and where weavers were fined for the smallest imperfections. (10-230)

After the intense strike of 1889, Filippelli notes that:

> Once again, the textile workers of Fall River had gone to war with their powerful employers; once again they had achieved a solidarity among the trades that was the envy of most unions of the day; once again they had overcome gender and ethnic differences to forge a militant united front with the working-class community; once again they had lost. (33-177)

However, the strikes still continued, either in response to poor working conditions or when wages were cut. A cut in wages of 22.5 percent in 1904 resulted in a strike that lasted for a full six months, when 30,000 persons were without work or income. During this strike, 7,000 persons left the city. (32-47)

After 40 years of struggle, the 10-hour day was finally achieved in 1879, although mill owners found various ways to undermine it. It wasn't until 90 years following the first Fall River strike, however, that the 48-hour work week became a reality in the 1920s. In succeeding years, other workplace and progressive social legislation was passed at the state and federal levels, due to the growing voting strength, organizational abilities and political power of unionized industrial workers.

Environmental Pollution, Working Conditions, and Public Health

In a 1920 bulletin titled "Preventable Death in the Cotton Manufacturing Industry," the U.S. Bureau of Labor Statistics studied the death hazard of the entire population of Fall River between the years 1908 and 1912. The report concluded that mill operatives were twice as likely to die from tuberculosis than non-mill workers and that the risk of death from other causes was 20 percent higher for mill workers than for other members of the population. The spinning and weaving rooms showed "the greatest peril for male workers" while, for females, the spinning room showed the most mortality, due in part to the more youthful female population in the spinning rooms. Tuberculosis accounted for 45 percent of all deaths of mill operatives aged 15 to 44 during the study period.

The study intimates that the pervasive lint in the cotton textile mills is highly correlated with the high rate of tuberculosis in mill workers. Observers have long noted that mill workers conducting certain operations look like "old ghosts" because of their covering of white lint at the end of the day.

The infant mortality in the mill towns was very high. In *Working Class Community in Industrial America*, John Cumbler observes that "Fall River earned the infamous distinction of having the highest infant death rate of any northern city, far surpassing New York City and even the mill cities of Lancashire, England." (20-252)

High humidity is a necessity in cotton textile mills for the proper creation of yarn and thread. However, this humidity can be oppressive to human operatives. In his 1902 account of a visit to Fall River's mills, T.M. Young, from England, relates that:

> In the mills the air is, as a rule, very bad, and there is often no provision at all for proper ventilation. In many mills I have seen the condensed moisture streaming down the window, and the clouds of water-vapour, almost scalding hot, rising amongst the looms from open grids on the floor. On the whole I should say that the conditions in which factory labour is performed, even at this temperate season [April] and in this model State, are very much more trying to human endurance and health than in Lancashire. (88-5)

Thomas Russell Smith mentions that, before the introduction of mechanical humidifiers, New England textile manufacturers attempted to replicate the humid conditions in the Lancashire mills by various means, including injecting steam directly into the weave rooms:

> ... with resulting unpleasantness and injury to the health of the operatives. This seems to have been the prevalent practice in Fall River, for in 1882 the operatives complained 'concerning the use of hot steam in weave rooms, where windows were kept closed in order that warp and filling might be kept damp. The heat was said to be insufferable in summer.' (73-58)

In the *Run of the Mill,* Dunwell mentions that:

> The close, humid, lint-filled air within the mills propagated pneumonia, tuberculosis, and other respiratory diseases. By the turn of the century, nearly seventy percent of all deaths among textile workers at Lawrence were caused by respiratory disease. One spinner in three died before completing 10 years of factory work. Half of these victims were under twenty-five years of age. (24-119)

The Immigrant Culture

Following the Civil War, northern mill owners faced severe labor shortages and, to meet this demand, recruiters began to solicit French Canada and the British Isles for new workers, promising steady work at high wages. This led to a new wave of immigration not only from Ireland but also from England, Scotland, and Canada. English textile workers were in high demand because of their knowledge of the newer textile machinery, since English engineering led in that area.

In the 1870s, French Canadians were the major immigrant group moving to Fall River, when the city was experiencing explosive growth. From 1870 to 1880, Fall River's French Canadian population grew from 500 to 11,000 and continued to grow in the following decades. By 1900, more than one-third of the population of Quebec had moved south to work in New England's mills.

fig. 14.6 **The Fall River dinner pail**

During the one and one-half hour lunch break from school, many Fall River children delivered dinner pails to fathers or other family members who worked in the textile mills along the Quequechan River and elsewhere in the city. As shown in the above photo, the dinner pails were made of mottled gray enameled steel and had two compartments, with the top slipping into the bottom compartment.

The dinner pail shown here is owned by Lionel (Leo) Cadrin. As a child, Mr. Cadrin brought his father, Amie, his lunch in this pail. Amie Cadrin worked in the Berkshire Hathaway Mills on Grinnell Street. Leo says that the lunches prepared by his mother always consisted of hot food such as stews, roasts, or fricassees. On Friday, lunch was clam chowder. Typically, a lunch bag carrying bread was tied to the handle of the pail. The top of the pail was usually tied down so that the swinging of the pail by the kids did not spill the contents.

Every day, Leo would walk the considerable distance from his home on Eastern Avenue to the Berkshire Hathaway Mills, where his father unloaded bales of cotton into a shed and sorted the cotton for quality. Leo would arrive about 11:40 and his father would put the pail on the hot steam pipes that passed through his work area to keep the food hot until his lunch break. As a young girl, Leo's wife Lorraine brought a dinner pail to her brother who worked at the Kerr Thread Mill.

Germaine Sirois and Joseph Jean remember that coffee was placed in the bottom of the pail and food in the upper compartment. The hot coffee kept the food warm.

Monsour (Raymond) Monsour, Violet Howayeck, and Alice Hassoun remember that, when they delivered dinner pails to their fathers, Lebanese families placed the main meal in the bottom of the pail and a chopped Lebanese salad in the upper compartment. A loaf of Syrian bread was also brought.

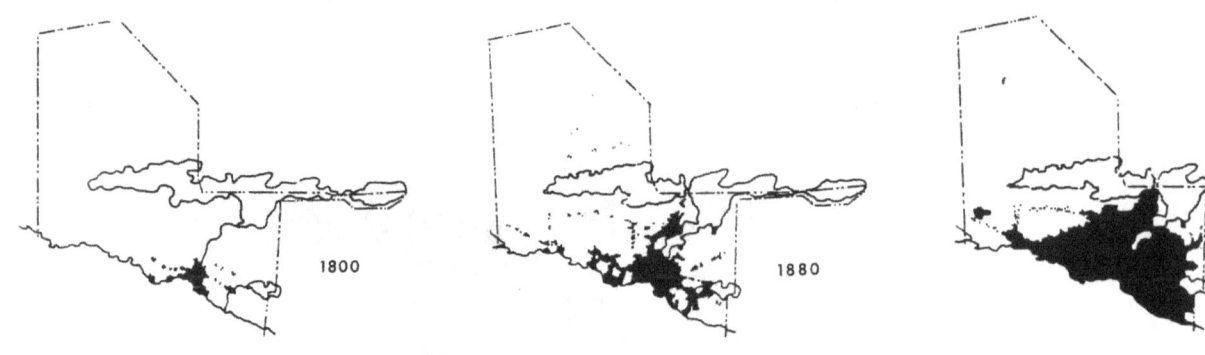

fig. 14.7 **The progression of urban growth in Fall River**

Note the successive growth of the city on either side of the Quequechan River

Source: *Fall River 2000*

In the mid 1870s, the Fall River work force was 25% American natives, 33% English (most from Lancashire), 20% Irish, and about 20% French Canadian.

The next great wave of immigration came from the Azores Islands, a legacy of the whaling era, when whaling ships would stop at the Azores to add to their crews. This Portuguese population settled mostly in Fall River and New Bedford but also in other mill towns in much smaller numbers.

Polish, Italian and Greek immigrants, Russian Jews, and Lebanese Christians followed. A wave of Italian immigration came as a result of the need for workers for the granite quarries. Skilled stonecutters were recruited by the Beattie quarry to meet the demand for a contract with the railroads, when grade separation required large amounts of granite for bridges, retaining walls and abutments. This Italian community developed around the Beattie quarry located at North Quarry and Beattie Streets.

As immigration continued, Fall River, Lawrence, Lowell, and Manchester became communities with the highest percentage of foreign-born population.

As Steve Dunwell notes in *The Run of the Mill:*

> Each successive national group made its way to the factory town to occupy the lowest ranks of the work force, replacing earlier immigrant workers who had already advanced to more skilled positions. Each nationality participated in turn, in an immigrant cycle leading from helpless destitution toward economic and social stability, from powerlessness to influence, and from alien status toward Americanization.

The Flint community adjoining the Quequechan River, which received its name from the Flint mill complex, is a good example of how neighborhoods developed in Fall River.

A RIVER AND ITS CITY

The first mills developed in the area were built in the mid-1870s, when the city saw a burst of mill building following the end of the Civil War and following the development of reliable steam power. Mill housing was built within walking distance along the streets adjoining the mills, and the city added schools and infrastructure.

When the immigrants arrived, their religious leaders provided churches for worship, parochial schools for instruction, and housing for the priests and the religious teaching

fig. 14.8 **Downtown Fall River, about 1890**

Source: David Symons in *Fall River 2000*

orders. Religious institutions and language were the principal means of preserving the ethnic identity, culture, and cohesion of each immigrant group. While the Roman Catholic hierarchy resisted the concept of ethnic churches, each ethnic group insisted on having its own church with priests that spoke their own language.

Each neighborhood included several Catholic churches, each affiliated with its own ethnic group. Every ethnic neighborhood in Fall River included at least three Catholic churches: one for the Irish, one for the French, and one for the Portuguese. In certain neighborhoods, other churches represented other Catholic and Orthodox immigrant groups—Polish, Italian, Lebanese, Greek, and Ukrainian. Protestant denominations included Episcopalian (English immigrants), Congregationalists, Quakers, Baptists, Methodists, Presbyterians, and Unitarians, among others.

In the Flint, French residents represented the majority of the population, and their church, Notre Dame de Lourdes, dominated the skyline with its physical presence. The Irish worshiped at Immaculate Conception Church and the Portuguese at Espirito Santo Church. The sizable population of Lebanese Marionite Catholics that settled in the Flint worshiped at St. Anthony of the Desert Church. The English attended St. Mark's Episcopal Church or churches of other Protestant denominations and the Jewish population in the neighborhood worshiped at their synagogue.

The French in particular were great institution builders, represented in the Flint by the Notre Dame de Lourdes complex. The church itself was massive and built of Fall River granite. The column-less interior of the church was without doubt the most beautiful and impressive interior in the city, an awe-inspiring, highly decorated space with stained glass windows from France and ceilings covered with paintings of religious scenes by Italian artists.

The complex included Notre Dame Elementary School, Jesus Mary Academy high school for girls, Prevost High School for boys, St. Joseph's Orphanage, a convent for the teaching nuns, housing for the teaching brothers, and a rectory for the many priests assigned to the parish. The only institution that it didn't provide was a Catholic hospital, and that was only because one was already being provided by the St. Anne's Church complex, affiliated with the French community in the South End of the city.

15 The Environmental Impact of Industrialization on the Quequechan River

Historical Sources of the River's Pollution

Before 1850, the upper Quequechan River was fairly pristine. Drinking water was taken from the river at the "Landing Place" on Hartwell Street, where the Electric Light Company is now located and sold by the pail and hogshead for domestic use. Water was also piped to dwellings below the dam. Before the introduction of city water in 1875, a "town pump" was maintained at the northwest corner of Pocasset and Second Streets.

Prior to the introduction of city water in 1875, water for fire protection for the mills on the falls and for property below the hill was provided by a pipe laid from the dam to the Iron Works property.

Virtually all of the pollution from households, factories, tanyards, and slaughterhouses at that time occurred just above and along the falls.

With the introduction of steam power after mid-century, however, the textile industry was freed from its locational constraints and mills quickly began being built above the falls on abundant flat land on either side of the river. Along with the mills were built housing for

Alfred J. Lima

Examples of pollution sources on the Quequechan River, 1915

Source: Fay, Spofford, and Thorndike: *Report of the Watuppa Ponds and Quequechan River Commission to the City Council, City of Fall River, 1915*

fig. 15.1
Privies overhanging river above Watuppa Dam, October 2, 1914

fig. 15.2
Choate Street sewer. Crude sewage discharged into river, October 2, 1914

fig. 15.3
Privies, south end of Blossom's Avenue, October 2, 1914

fig. 15.4
3:15 pm, September 22, 1909, from 100 feet below Fall River Laundry, looking down stream

fig. 15.5
Privy at discharge from Hargraves Mills Nos. 2 and 3, October 2, 1914

fig. 15.6
City dump between Lawrence and Salem Streets, October 13, 1914

workers, and neighborhoods quickly sprang up along either side of the river. As the river gradually became polluted, it no longer became visible for domestic purposes.

A national outbreak of cholera in 1854 hit hardest the poor Irish immigrants in Fall River who were blamed for their misfortunes by local anti-immigrant Know-Nothings, who labeled the disease "rum cholera." As the contagion spread through the city's water sources, a citizen's committee resolved to halt the spread of the disease by ordering the Board of Health to destroy all liquor owned in the city. Only the death of a prominent Yankee woman from cholera stayed the action.

In 1857, Mayor Nathaniel B. Borden insisted that a sewage system be built, but municipal officials objected that the cost of blasting through the ever-present ledge in the city would be prohibitive. When sewers were finally constructed, they were combined stormwater and sewer systems with outlets to the Quequechan and Taunton Rivers and Mount Hope Bay.

Past Efforts to Remediate Pollution in the River

In 1915, Fay, Spofford, and Thorndike in their report to the city stated flatly "the Quequechan River is an open sewer. It is a detriment to its immediate vicinity and, through the possibility of spreading disease, is a menace as well to the whole city." As late as 1915, open trenches of sewage flowed into the river from neighboring residential areas. Outhouses attached to the mills and businesses along the stream were cantilevered over the river. Over time, this sewage, mill waste, and street washings built up on the bottom of the slow-moving river above the dam.

In addition to organic pollution, the river suffered from thermal pollution from the textile mills. Each mill had an intake from the river that took in cold water into its condensers to cool its steam engines, and each mill had an outflow that exited hot water from its condensers. Eventually, the river became so hot, particularly in the summer-time, that it became ineffective as cooling water.

As early as 1875, Phineas Ball, a Worcester civil engineer commissioned to study the city's sewage system, laid special stress on correcting the pollution of the Quequechan River. By 1883, the situation had become so serious that the state legislature passed a special act addressing the issue. However, no corrective action was taken by the city.

In 1915, the Watuppa Ponds and Quequechan River Commission prepared a report with engineers Fay, Spofford, and Thorndike that proposed a three-part solution to the problem. The solution proposed by the engineers was ingenious. They proposed a three-part conduit, one part to pipe cold water from South Watuppa Pond directly to the mills. Another part of the conduit would bring the hot water back to a holding basin at the mouth of the Quequechan which, when cooled, would flow into the South Watuppa Pond. A third part of the conduit would carry storm water away from the city to Mount Hope Bay. None of that plan was implemented.

In 1954, another study of the Quequechan River was conducted by the engineering firm of Hayden, Harding, and Buchanan. That study said that:

> The Quequechan River at the present time is a grossly polluted, foul smelling, slow flowing stream whose existing bad condition is due in part to present-day sewage overflow, street washings, and to over a century of accumulated sludge deposits from sewer overflows, cesspool overflows, and mill wastes which have settled on the flats and have formed a rich organic bottom. This organic matter stimulates vegetation, and when uncovered gives off offensive odors. The original river bed is obscured now by this thick layer of sludge, and the slow flow of water through the shallow flats makes any scouring action impossible ... Examination of bacteriological data [that we have collected] substantiates our statement that the Quequechan River is a grossly polluted stream ... It is our opinion that the quality of the water in the Quequechan River is unsuitable for anything but the roughest industrial usage, and furthermore is an ever-present heath menace to the inhabitants of the City of Fall River. (43-10)

The Hayden, Harding, and Buchanan report proposed an open channel from the beginning of the Quequechan River at the "sand bar" to the Third Street dam in order to provide up to 30 million gallons of clean processing water for industrial use. That plan was also not implemented.

The City of Fall River's sewer system currently consists of 176 miles of sanitary and combined stormwater and wastwater sewers that serves approximately 88,000 residents and many industries. Approximately 75% of this system is combined wastewater and stormwater sewers. Providing separate lines for wastewater and stormwater was not seriously implemented by the city until the 1960s. Before the 1940s, when the city's primary sewage treatment plant was built, the combined sewers emptied raw sewage into the Quequechan River, the Taunton River, or Mount Hope Bay. The combined sewers still flow into these waters during rainstorms, when the volume of water overtaxes the wastewater treatment plant.

Today, the Quequechan River flows as it always has, with its same seasonal fluctuations and rhythms. While the river has been forgotten for generations and has been shunned because of its foul condition, new initiatives are occurring that promise to make the river a unifying element and an asset to residents.

Current Efforts to Clean Up the River

The most significant change in the river in many decades is the current effort by the City of Fall River to clean up the river through its Combined Sewer Outflow (CSO) initiative. This initiative began with a suit brought by the Conservation Law Foundation under the provisions of the federal Clean Water Act that resulted in a court order to correct sources of pollution of Mount Hope Bay and the Taunton River. Planning for this project began in the early 1980s. This initiative involves two main phases, one to increase the capacity of the wastewater treatment plant to handle larger storms and the second to build a storage capacity to hold the larger volumes of sewage and stormwater that results from major storms.

Phase one involves upgrading the wastewater treatment plant on Bay Street to increase its flow capacity, resulting in fewer CSO discharges during light rain events. This upgrade will increase the plant's capacity from 50 million gallons a day of sewage capacity to 75 million gallons a day. This capacity increase is necessary to allow the plant to reliably treat the increased combined drainage/sewage flow that will result from the storage of flows in the storage tunnel. This first phase was constructed between 1997 and 2000 at a cost of $23 million.

Phase two involves the construction of a deep rock storage tunnel that is seven miles long and 20 feet in diameter. This storage tunnel will collect and store sewage and stormwater runoff from large storms, store the water, and release it gradually to the treatment plant for processing. In addition, the tunnel will act as an interceptor for 80% of dry weather flow, relieving the existing system. The tunnel was completed in 2004. The combined cost of phases one and two totaled $98 million.

Stage three will implement drainage improvements between the tunnel and the existing main interceptors, including the separation of sewers from drain lines. (92)

16 The Future of the Quequechan River

The construction of Interstate I-195 in the middle of the Quequechan River exposed the city's worst side and transformed the river into the city's front door. As drivers passed along the highway, what they saw was the back side of mill complexes, the city incinerator, an abandoned railway bed and discarded debris of various kinds. It was not a pretty picture and still remains unsightly.

Worse, since the construction of the highway, the appearance of the city from the highway has deteriorated visually because of the proliferation of billboards and signs. In addition, the removal of windows in many of the granite textile mills, the filling in of the spaces with concrete block, and the removal of roof brackets have transformed once-majestic buildings into unattractive blocks. The overall effect is a vista of ugliness that detracts from the city's image and only reinforces the negative opinion of Fall River.

If Fall River wants to do something about its image, the place to start is at its front door, the Quequechan River.

Unfortunately, there has never been a plan to address the many issues related to the Quequechan.

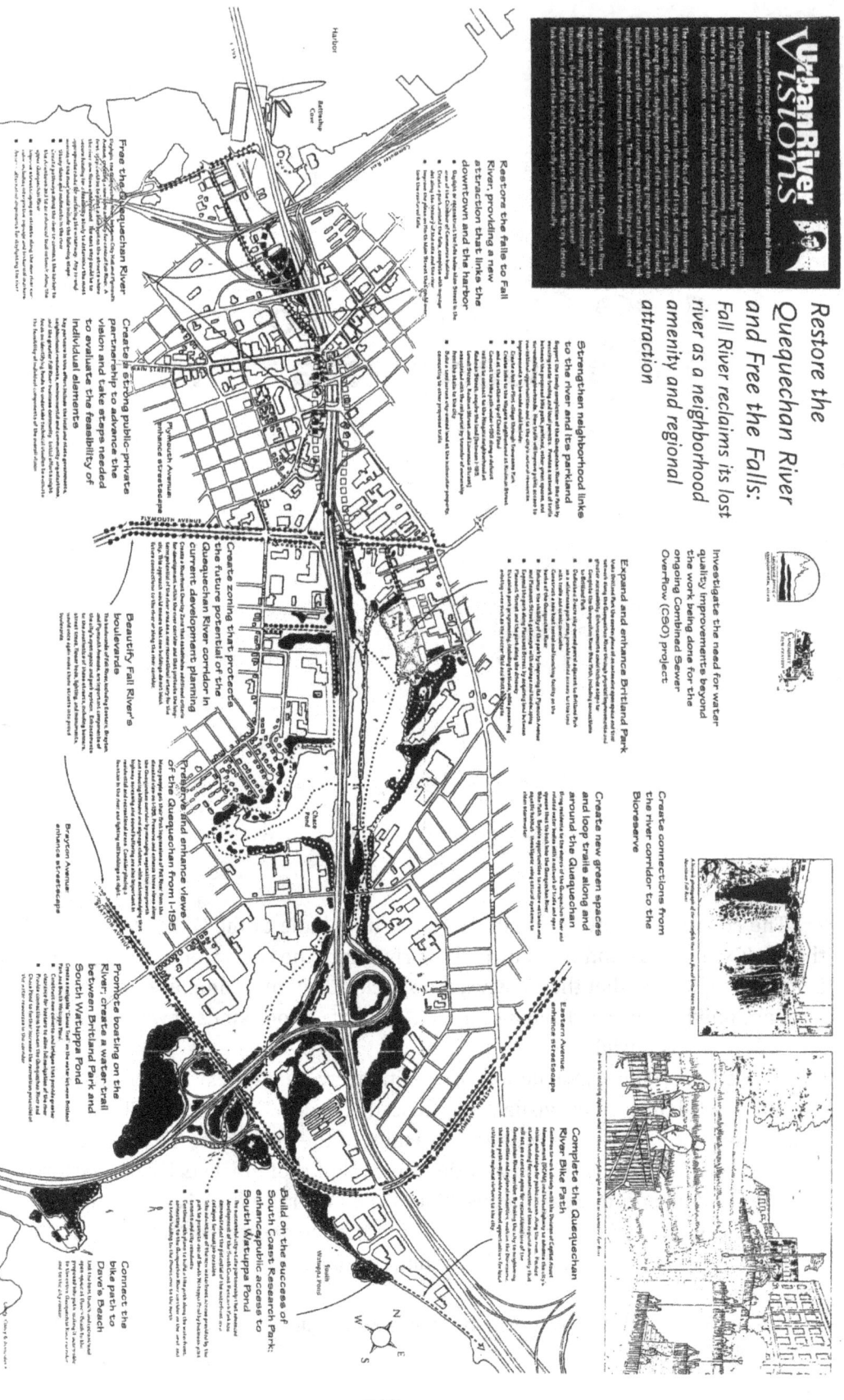

fig. 16.1 **Recommendations of the Fall River Urban River Visions Initiative**

Poster graphic: Executive Office of Environmental Affairs

During a forum on the Quequechan River, sponsored by Green Futures and held in 2001, the idea for daylighting the river was first raised. John Warner, a leader in the daylighting of the rivers in downtown Providence, described how the idea there grew from a vague vision to a reality and the impact that this project has had on the revitalization of Providence.

That forum was attended by representatives of the Executive Office of Environmental Affairs, and it became the catalyst for the creation of EOEA's Urban River Visions initiative. The Urban River Visions initiative funded the development of planning efforts to advance the use of urban rivers as a focus for community revitalization in seven communities in the Commonwealth, including the Quequechan River in Fall River. EOEA retained the planning and design consulting firm of Goody, Clancy, and Associates to convene the forums and prepare graphic materials and compile recommendations. A public forum was held on November 23, 2002, to bring Fall River residents together and to discuss ideas for the future of the river. The concepts that resulted from the initiative represent a comprehensive and bold plan for the renewal of the river and its use as a focus for urban revitalization. The text of the Fall River Urban River Visions plan follows:

The Quequechan River and the waterfalls that once graced the western part of Fall River gave the city its name and its identity. They provided the power for the mills that once drove the city's economy. Today, however, the river's potential as an amenity has been obscured by the impacts of highway construction, contaminated sediments, and sewer overflows.

The community's vision centers on the idea of restoring the river and making it visible once again, freeing it from the shadow of I-195, and improving water quality. Important elements of the vision include completing a bike path along the river, daylighting portions of the river that are now buried, restoring the falls below Main Street, developing programs and signage to build awareness of the river, and creating new parkland and trails that link neighborhoods and natural areas. The technical feasibility and costs of implementing each element of this vision should be explored.

As the river is restored, the dramatic waterfall on the Quequechan River can again become Fall River's defining natural feature. Now hidden under highway ramps, enclosed in a pipe, and threaded through historic mill structures, the path of the Quequechan has long been obscured. Restoration of the falls could be the catalyst that fulfills the city's desire to link downtown and the harbor, physically and economically.

The specific recommendations of the Urban River Visions plan include:

Restore the falls to Fall River; providing a new attraction that links the downtown and the harbor
- Daylight or reconstruct the falls below Main Street in the area of the Chamber of Commerce building.
- Create a park around the falls, complete with signage detailing the history of the mills and the river.

- Improve the plaza on North Main Street that could overlook the restored falls.

Free the Quequechan River

Daylight the Quequechan River between City Hall and Plymouth Avenue, creating a new civic amenity for central Fall River. A first step could be to paint a blue trail on the streets where the river now flows underground. The next step could be to secure funding for a feasibility study to determine the most appropriate route for surfacing the waterway. Any re-engineering of the river should include the following steps:

- Study flows and sediments in the river.
- Create pathways along the river to connect the harbor to the downtown and to an enhanced trail network along the upper Quequechan River.
- Improve streetscaping on streets along the new river corridor, including interpretive signage and historical markers.
- Assess alternative alignments for daylighting the river.

Complete the Quequechan River Bike Path

Continue to work closely with the Division of Capital Asset Management (DCAM) and MassHighway to advance the city's vision and design for public access along the river. Finalize state funding for construction of this regional amenity that will act as a central spine for recreational use of the Quequechan River corridor. By linking the city to neighboring communities and regional amenities, such as the Bioreserve, the bike path will provide recreational opportunities for local citizens and regional visitors to the city.

Create new green spaces and loop trails along and around the Quequechan

Bring residents to the shores of the Quequechan River and related water bodies with a network of trails and open spaces that tie back into the Quequechan River Bike Path. Explore opportunities to restore wetlands and aquatic habitat. Investigate using natural systems to clean stormwater.

Expand and enhance Britland Park

Make Britland Park the centerpiece of an extensive open space and trail network along the Quequechan River through physical improvements and greater accessibility. Enhancements could include steps to:

- Complete the Quequechan River Bike Path, including connections to Britland Park.
- Dedicate a 2-acre city-owned parcel adjacent to Britland Park as a wilderness park area; provide limited access to this land with trails and scenic overlooks.
- Construct a new boat rental and launching facility on the banks of the Quequechan River.

- Enhance the visibility of the park by improving its Plymouth Avenue and Pleasant Street gateways with signage and landscaping.
- Expand the park along Pleasant Street by acquiring land between Pleasant Street and the park along the driveway.
- Establish park programming, including festivals, while preserving existing uses such as the soccer field and BMX bike trails.

Strengthen neighborhood links to the river and its parkland

Support the timely completion of the Quequechan River Bike Path by securing state funding and local permits. Provide a network of trails between the proposed bike path, parkland, other green spaces, and surrounding neighborhoods. New trails will improve public access to recreational opportunities and to the city's natural resources. Improvements to be made could include:

- Create a link to Flint Village through Father Travassos Park.
- Create links to the Niagara neighborhood at Rodman Street and at the southern tip of Chace Pond.
- Connect the bike path under I-195 along a defunct rail line to connect to the Niagara neighborhood at Rodman Street; acquire the land (between I-195, Lowell Street, Rodman Street and Lawrence Street) associated with the rail parcel by transfer of ownership from the state to the city.
- Build a trail across city-owned land at the incinerator property, connecting to other proposed trails.

Create connections from the river corridor to the Bioreserve

Preserve and enhance views of the Quequechan from I-195

Many people get their first impressions of Fall River from the elevated view on I-195. Preserve and enhance these views along the Qeuquechan corridor by managing vegetation overgrowth and reducing billboard and signage clutter, while acknowledging that highway screening and sound buffering are also important in residential and recreational areas. Consider placing a fountain in the river and lighting mill buildings at night.

Beautify Fall River's boulevards

The boulevards of Fall River, including Eastern, Brayton, and Plymouth Avenues, are important components of the city's open space and park system. Enhancements to the aesthetics of these streets—including banners, street trees, flower beds, lighting, and monuments—would once again make these streets into proud boulevards.

Promote boating on the river; create a water trail between Britland Park and South Watuppa Pond.

Create a navigable "Canoe Trail" on the water between Britland Park and South Watuppa Pond.

- Construct new culverts and bridges that provide greater clearance for boaters to allow full navigation of the river.
- Provide connections between the Quequechan River and Chace Pond to further increase the recreation potential of the water resources in the corridor.

Create zoning that protects the future potential of the Queqeuchan River corridor in current development planning.

Create a Riverfront Overlay Zone that establishes additional criteria for development within the river corridor and that protects the long-term potential of the river area as a new recreational artery for the city. This approach would ensure that new buildings do not block future connections to the river or along the river corridor.

Build on the success of South Coast Research Park; enhance public access to South Watuppa Pond

- The successful city-state partnership that advanced development of the South Coast Research Park has demonstrated the potential of the waterfront as a catalyst for local job creation.
- Take advantage of the new waterfront access provided by the park to promote use of South Watuppa Pond by business park tenants and city residents.
- Continue with plans to build a bike path along the waterfront, connection to the Quequechan River corridor on the west and to the trails leading to the Bioreserve to the north.

Connect the bike path to Dave's Beach

Link the boat launch and recreational open space at Dave's Beach to the proposed bike path, making it accessible to the entire Quequechan River corridor and to the city center.

Investigate the need for water quality improvements beyond the work being done for the ongoing Combined Sewer Outflow (CSO) project

Create a strong public-private partnership to advance the vision and take steps needed to evaluate the feasibility of individual elements

Key partners in this effort include the local and state governments, neighborhood residents, environmental and community organizations, and the greater Fall River business community. Initial efforts might focus on identifying funds to undertake technical studies to evaluate the feasibility of individual components of the overall vision.

Implementing the Urban River Visions plan

The City of Fall River has taken a major lead in implementing the Urban River Visions plan through the implementation of the Combined Sewer Outflow project, a massive and costly public works construction program. This CSO initiative will result in major improvements in the water quality of the Quequechan River.

The city is also pursuing the realization of the Quequechan River bike path. The first phase of the bike path began construction in 2006, and extends from the Westport town line to Brayton Avenue. Construction of the next phase of the bikepath, from Brayton Avenue to Plymouth Avenue, will begin 2014.

The Quequechan River provides an excellent opportunity for using the city's history as a vehicle for building a new future. Visually, Interstate I-195 is Fall River's Main Street, the view from which others see the city. Transforming that vista requires a combination of environmental protection and historic preservation. The Quequechan River Greenway can serve as a framework for this effort.

For this to happen, the recommendations of the Urban River Visions initiative need to be made more concrete by taking each recommendation and identifying what needs to be done in each of the next several years to realize the revitalization of the river valley. However, visions do not implement themselves. Visions come to life when advocates work to make them happen.

"If you build castles in the sky," Henry David Thoreau once said, "your work need not be lost. That is where they should be. Now put the foundations under them."

References

1. Albion, Robert G., William A. Baker, Benjamin W. Labaree. *New England and the Sea.* Middletown, Connecticut: Wesleyan University Press, 1972.

2. American Printing Company. "Fabrics we are printing and how we do it." Fall River, MA, undated.

3. Angerstein, R. R. *R.R. Angerstein's Illustrated Travel Diary, 1753-1755: Industry in England and Wales from a Swedish Perspective.* Science Museum, 2001.

4. Anonymous. *History of the Town of Freetown, Massachusetts.* Fall River, Massachusetts: J.H. Franklin and Company, 1902.

5. Anonymous. "The First Top Forge at Wortley, South Yorkshire." From the Internet.

6. Anonymous/unknown. "Industrial Art" in *The International Exhibition.* Philadelphia, 1876.

7. Athearn, Roy, Arthur Staples and Carol Barnes. "A Middle/Late Shell Midden at Peace Haven 2." *Bulletin of the Massachusetts Archaeological Society.* Boston: Massachusetts Archaeological Society, Inc., October, 1972-January, 1973.

8. Athearn, Roy, Arthur Staples and Carol Barnes. "Peace Haven 2: M39-74." *Widening Horizons.* Boston: Massachusetts Archaeological Society, April 12, 1980.

8a. Baines, Sir Edward. *History of the cotton manufacture in Great Britain; with a notice of it early history in the East, and in all quarters of the globe.* London: H. Fisher, R. Fisher & P. Jackson, 1835.

9. Bidwell, Percy Wells and John I. Falconer. *History of Agriculture in the Northern United States, 1620-1860.* Washington: Carnegie Institution of Washington, 1925.

10. Blewett, Mary H. *Constant Turmoil: The Politics of Industrial Life In Nineteenth Century New England.* Amherst, Massachusetts: University of Massachusetts Press, 2000.

11. Bordewich, Fergus M. *Bound for Canaan: The Underground Railroad and the War for the Soul of America.* New York: HarperCollins Publishers, Inc., 2005.

12. Boyer, Richard O. and Herbert M. Morais. *Labor's Untold Story*. New York: United Electrical, Radio and Machine Workers of America, 1955.

13. Bradbury, James E. *Seafarers of Somerset*. Somerset, Massachusetts, 1996.

14. Brayton, Roswell. *The Cause of the Industrial Decline of Fall River and the Effects of the Decline Upon the City*. Cambridge, Massachusetts: Harvard College, 1939.

15. Brown, F.K. *Through the Mill*. Boston: Pilgrim Press, 1911.

16. Brown, Richard D. *Massachusetts: A Bicentennial History*. New York: W. W. Norton and Company, 1978.

17. Burrows, Fredrika A. *Cannonballs and Cranberries*. Taunton, Massachusetts: William S. Sullwold, Publishing, 1976.

18. Conforti, William J. *Fall River's Watuppa Reservation: A Brief Account of Its Origins and Evolution*. Fall River, Massachusetts, 1996.

19. Cronon, William. *Changes in the Land: Indians, Colonists and the Ecology of New England*. New York: Farrar, Straus and Giroux, 1983.

20. Cumbler, John T. *Working Class Community in Industrial America: Work, Leisure and Struggle in Two Industrial Cities, 1880-1930*. Westport, Connecticut: Greenwood Press, 1979.

21. Curley, John R., et al. *A Study of Marine Resources of the Taunton River and Mount Hope Bay*. Boston: Executive Office of Environmental Affairs, August, 1974.

22. Dow, George Francis. *Everyday Life in the Massachusetts Bay Colony*. New York: Dover Publications, 1988. Originally published in 1935 by The Society for the Preservation of New England Antiquities.

23. Drake, James D. *King Philip's War: Civil War in New England, 1675-1676*. Amherst, Massachusetts: University of Massachusetts Press, 1999.

24. Dunwell, Steve. *The Run of the Mill: A Pictorial Narrative of the Expansion, Dominion, Decline and Enduring Impact of the New England Textile Industry*. Boston: David R. Godine Press, 1978.

25. Earl, Henry H. *A Centennial History of Fall River, Massachusetts*. New York: Atlantic Publishing and Engraving Company, 1877.

25a. Eayrs, Frederick, Jr. "The Iron Manufacture at Judge Oliver's Works" *The Middleborough Antiquarian*: Vol. VII, Number 1, February, 1965.

26. Emery, Samuel Hopkins, D.D. *History of Taunton, Massachusetts*. Syracuse, New York: D. Mason and Company, Publishers, 1893.

27. Fall River, City of. *Fall River 2000: An Overall Development Plan for Fall River*. Fall River: Fall River Planning Department, 1966.

28. Fall River, City of. *Survey of Historical Structures*. Fall River: Office of Historic Preservation, 1981.

29. Fall River, City of. *Fall River Open Space and Recreation Plan*. Fall River, Massachusetts: Open Space and Recreation Committee, 1997.

29a. *Fall River News* and *The Taunton Gazette. Our County and Its People: A Descriptive and Biographical Record of Bristol County, Massachusetts*. Boston: The Boston History Company, 1899.

30. Fay, Spofford, and Thorndike. *Report of the Watuppa Ponds and Quequechan River Commission to the City Council, City of Fall River*. Boston: Fort Hill Press, 1915.

31. Fenner, Henry M. *History of Fall River*. New York: F.T. Smiley Publishing Co., 1906.

32. Fenner, Henry M. *History of Fall River, Massachusetts*. Fall River, Massachusetts: Fall River Merchants Association, 1911.

33. Filippelli, Ronald, editor. *Labor Conflict in the United States: An Encyclopedia*. New York: Garland Publishing, Inc., 1990.

34. Fisher, David Hackett. *Albion's Seed: Four British Folkways in America*. New York: Oxford University Press, 1989.

35. Fisher, Leonard Everett. *The Blacksmiths*. New York: Benchmark Books, 1976.

36. Fletcher, Sir Banister. *A History of Architecture*. New York: Charles Scribner's Sons, 1975. 18th edition revised by J.C. Palmer.

37. Foster, George H. and Peter C. Weiglin. *Splendor Sailed the Sound: The New Haven Railroad and the Fall River Line*. San Mateo, CA: Potentials Group, Inc., 1989.

38. Fowler, Rev. Orin. *History of Fall River, with Notices of Freetown and Tiverton*. Fall River, Massachusetts: Almy and Milne, Printers, Daily News Steam Press, 1862 (originally published in 1841).

39. French, B. F. *History of the Rise and Progress of the Iron Trade of the United States, from 1621 to 1857*. New York: Wiley and Halsted, 1858.

40. Gardner, J. Starkie. *Ironwork*. London: Chapman and Hall, Ld., 1893.

41. Hart, William A. *History of the Town of Somerset, Massachusetts*. Somerset, Massachusetts: Town of Somerset, 1940.

42. Hartley, Edward N. *Ironworks on the Saugus*. Norman: University of Oklahoma Press, 1957.

43. Hayden, Harding and Buchanan. *Investigations and Report on Quequechan River, City of Fall River, Massachusetts*. Boston, 1954

44. Howe, George. *Mount Hope: A New England Chronicle*. New York: The Viking Press, 1959.

45. Hurstwic. "Iron Production in the Norse Era." From the Internet.

46. Johnson, Eric S. and Thomas F. Mahlstedt. "The Roy Athearn Collection: A Report from the Massachusetts Historical Commission's Statewide Survey." *Bulletin of the Massachusetts Archaeological Society*: April, 1984.

47. Johnson, Steven F. *Ninnouck: The Algonkian People of New England*. Marlborough, Massachusetts: Bliss Publishing Company, 1995.

48. Juravich, Tom, et al. *Commonwealth of Toil: Chapters in the History of Massachusetts Workers and their Unions*. Amherst, Massachusetts: University of Massachusetts Press, 1996.

49. Kurlansky, Mark. *Salt: A World History*. New York: Walker Publishing Company, 2002.

50. Lane, Helen S. *History of the Town of Dighton, Massachusetts*. Dighton, Massachusetts: The Town of Dighton, 1962.

51 Lima, Alfred J. *The Taunton Heritage River Guide: A Guide to the Appreciation of the Taunton River in Fall River*. Fall River, Massachusetts, 2002.

52. Lintner, Sylvia Chace. "Mill Architecture in Fall River 1865-1880." Reprinted from *The New England Quarterly:* Vol. XXI. No. 2 (June, 1948).

53. Linton, George E. *Natural and Manmade Textile Fibers: Raw Material to Finished Product*. New York: Duell, Sloan and Pearce, 1966.

54. Lovell, Malcolm R. *Two Quaker Sisters*. New York: Liveright Publishing Corporation, 1937.

55. Macaulay, David. *Mill.* Boston, Massachusetts: Houghton, Mifflin Co., 1983.

56. Massachusetts Foundation for the Humanities. Mass Moments. "Explorer Proves Wreck is Whydah." October 30, 2005.

57. *Massachusetts Historical Commission. Historic and Archaeological Resources of Southeast Massachusetts*. Boston: Massachusetts Secretary of State, 1982.

58. Maiocco, Carmen. *Up the Flint.* Fall River, Massachusetts, undated.

59. McAdam, Roger Williams. *The Old Fall River Line*. New York: Stephen Daye Press, 1955.

60. Moore, Jim. "Some Balanced Rocks Across Massachusetts." *New England Antiquities Research*. From the Internet, 2005.

61. Munro, Wilfred H. *Mount Hope Lands: From the Visit of the Northmen to the Present Time.* Providence: J.A. & R.A. Reid, Printers and Publishers, 1880.

61a. Murdock, William Bartlett. *Blast Furnaces of Carver, Massachusetts*. Poughkeepsie, NY, 1937.

62. Pearce, Benjamin Wood. *Recollections of a Long and Busy Life, 1819-1890*. Newport: B.W. Pearce, Newport Enterprise, 1890.

62a. Pfeiffer, Sacha. "N.H. sawmill is sharp reminder of the past." *The Boston Globe*: December 13, 2006.

63. Phillips, Arthur S. *The Phillips History of Fall River*. Fall River, Massachusetts: Dover Press, 1941. Fascicle I.

64. Phillips, Arthur S. *The Phillips History of Fall River*. Fall River, Massachusetts: Dover Press, 1941. Fascicle II.

65. Phillips, Arthur S. *The Phillips History of Fall River*. Fall River, Massachusetts: Dover Press, 1941. Fascicle III.

66. Prairie, Conner. "Fuel for the Fires: Charcoal Making in Nineteenth Century America." *Chronicle of the Early American Industries Association*: June, 1994.

67. Rigby, George Oliver. *Steep Brook*. Danvers, Massachusetts: Bett's Press, 1979.

68. Russell, Howard S. *A Long, Deep Furrow: Three Centuries of Farming in New England*. Hanover, New Hampshire: University Press of New England, 1976.

69. Russell, Howard S. *Indian New England Before the Mayflower*. Hanover, N.H.: University Press of New England, 1980.

70. Schultz, Eric B. and Michael J. Tougias. *King Philip's War: The History and Legacy of America's Forgotten Conflict*. Woodstock, Vermont: The Countryman Press, 1999.

71. Siebert, Wilbur H. *The Underground Railroad in Massachusetts*. Worcester, Massachusetts: American Antiquarian Society, 1936.

72. Silvia, Philip T. *The Spindle City: Labor, Politics and Religion in Fall River, Massachusetts, 1870-1905*. Fordham University thesis, 1973.

73. Smith, Thomas Russell. *The Cotton Textile Industry of Fall River*. Cambridge, Massachusetts: Harvard University Press, 1947.

74. Somm, Albert H. *Early American Wrought Iron*. New York: Bonanza Books, 1989.

75. Speed, John Gilmer. *A Fall River Incident, or a little visit to a big mill*. Fall River: American Printing Company, 1895.

76. State of New Jersey. "Iron Plantations in Jersey," from *Electronic New Jersey: A Digital Archive of new Jersey History*. New Jersey Core Content Curriculum Standards for Social Studies, 2006

77. Thompson, Charles O.F. *Sketches of Old Bristol*. Providence, Rhode Island: Roger Williams Press, 1942.

78. Thoreau, Henry David. "Slavery in Massachusetts," in *Civil Disobedience and Other Essays*. New York, NY: Dover Publications, 1993.

79. Tunis, Edwin. *Colonial Craftsmen and the Beginnings of American Industry.* New York, NY: The World Publishing Co., 1965.

80. Undocumented source. "Cotton mill health conditions." Undocumented news article dated January 6, 1920, in the Fall River Mills drawer in the Fall River Public Library.

81. Walton, Perry. *The Story of Textiles.* Boston, Massachusetts: John S. Lawrence, 1912.

82. *Wikipedia, the Internet encyclopedia.* Subject: Charcoal.

83. *Wikipedia, the Internet encyclopedia.* Subject: Millwright.

84. *Wikipedia, the Internet encyclopedia.* Subject: Pig iron.

85. *Wikipedia, the Internet encyclopedia.* Subject: Steel.

85a. *Wikipedia, the Internet encyclopedia.* Subject: Textile printing.

86. Wilbur, C. Keith. *The New England Indians.* Chester, Connecticut: *The Globe Pequot Press*, 1978.

87. Williamson, Scott Graham. *The American Craftsman.* New York: Crown Publishers, 1940.

88. Young, T. M. *The American Cotton Industry.* London, England: Methuen and Co., 1902.

89. Interview with Fernand Charles E. Auclair held on October 27, 2005.

90. Interview with Andrew Roy held on May 26, 2005.

91. Interview with Lionel and Lorraine Cadrin held on July 22, 2005.

92. Interview with and information provided by Fall River DPW Commissioner Terrance Sullivan.

www.ingramcontent.com/pod-product-compliance
Lightning Source LLC
Chambersburg PA
CBHW080411170426
43194CB00015B/2778